THE CLIMATE CHANGE SOLUTION

COMPLETE MITIGATION SCIENCE

Timothy C. Thompson

Copyright 2023 by Vestibule Holdings, Inc

All rights reserved. No part of this book may be reproduced or used in any manner without written permission of the copyright owner except for the use of quotations in a book review.

Edited by David Felstul and Kenneth Zink
Cover and Book Design by Veronica Scott

Credits: Cover Art, Shutterstock (ESB Professional)

ISBN: 979-8-218-96824-3 (Hardcover)
ISBN: 979-8-218-96823-6 (Paperback)
ISBN: 979-8-218-96825-0 (Ebook)

For more information: **www.ewc.company**

To my wife, you do deserve better, but you're stuck with me.

EARTH'S TO-DO LIST:

1. ~~Reducing Emissions Will Cure Climate Change!~~
2. ~~Solar Panels Will Cure Climate Change!~~
3. ~~Battery-Powered Cars Will Cure Climate Change!~~
4. ~~Wind Power Will Cure Climate Change!~~
5. ~~Nuclear Power, That'll Fix It!~~
6. ~~Tiny Houses! We'll All Live Like Mice!~~
7. ~~Reduce, Reuse, and Recycle Will Cure Climate Change!~~
8. ~~Planting Trees Will Cure Climate Change!~~
9. ~~Cap and Trade Will Cure Climate Change!~~
10. ~~Technology! We'll Invent Something to Cure It!~~
11. ~~Humans Can All Go Back to Being Hunter-Gatherers!~~
12. ~~Let's Become Vegans, That'll Cure Climate Change!~~
13. ~~Off-Grid Living Will Cure Climate Change!~~
14. ~~Our Government Will Cure Climate Change!~~
15. ~~Climate Change Isn't Real!~~
16. Complete Mitigation Science? Finally, a cure for climate change, really!

To the reader,

This book intends to update the reader's perception of climate change. It introduces "Complete Mitigation Science, CMS." CMS is possibly the most important climate development of this era because CMS is the only way to defeat CO2-driven climate change. That's a big claim, I know, I didn't believe it either when CMS revealed. It took some time for the concepts to absorb and my brain to adjust to an entirely new understanding of a real mother of a problem. It did eventually shape my opinion and convince me of advocacy except now I wish I had never heard CMS's message but eternally grateful I did.

As you read, you'll discover CMS the same way I did. We'll begin with how CMS developed, what CMS is, and explain how all this is especially important to everyone. Oh, and who I am as well.

Before reading the technical notes or published manuscripts, I suggest this book, which is not written in technically challenging jargon and is without the pomp and circumstance of academia, well sort of. The subject is sciency so unavoidable vernacular is sometimes used but avoided when I could.

I also must warn you. I might offend by making fun of pre-CMS climate beliefs. You can count on that. You can also count on not believing a word I write occasionally, but you will, eventually. It takes time to soak in new knowledge and CMS is that and some. And unfortunately, I'm known for truth for truth's sake and for applying higher-than-average accountability standards to everything that reeks of scientific disregard or emotional attachment to today's cult like misinformation followers. To be sure, I'm neither buying nor selling anything that's not immersed in relevant facts or significance. Which makes me abrasive to folks who refuse to be informed by facts, because facts only confuse them more.

So informally, that also makes me and CMS somewhat ugly to understand with your first glance. But, formally, we're just ugly enough to understand logically and therefore we become gorgeous with credibility. We both grow on you as you read, I promise. I can make and follow through with promises because logic does the messaging, but the study proofs provide all the glue needed to form a very tight bond with truth.

Finally, have no worries, there won't be a surprise test after this reading assignment. The only test left is if humans can survive their newly defined and unintentional contribution to climate change, a test that is now and unfortunately, pass or fail.

-*Timothy C. Thompson*-

CONTENTS

- ACKNOWLEDGEMENTS: 1
- LET'S BEGIN WITH REVENGE 2
- THE END OF CLIMATE CHANGE? 3
- PEOPLE PERSON? NO, AND IT'S NOT YOUR FAULT 35
- JUST ANOTHER CLIMATE MOUTHPIECE WITH A TREE 39
  - THE STUDY ONLY IMPROVED FROM THERE. 49
  - XPRIZE CARBON REMOVAL COMPETITION, BUT FIRST CHEERLEADER, NOT ME. 57
  - WHERE THE HECK WAS I? 63
- TO START, TRY EXPLAINING "IT" TO ANYONE 66
  - WHERE DID THOSE CARBON ATOMS FOUND WITHIN THE DRY WEIGHT OF TREES AND EVEN FOSSIL FUELS COME FROM? AS MENTIONED, PLANT LIFE. BUT THERE IS MORE TO TELL, GROWTH. 66
  - ODE TO THE LOGIC OF IT ALL. 74
  - SO, WHAT IS THE LOGIC BEHIND IT? OKAY, HOW ABOUT WHERE IT CAME FROM FIRST. 75
  - NOT THE WIN I'D HOPED FOR. SOMETIMES YOU GET FEATHERS WITH YOUR CHICKEN. 77
- BUT WILL HUMANS USE IT? 80
- INTERNAL PSYCHOBABBLE: PERCEPTION 83
- TIME TO TEST PERCEPTION 86
  - WHAT DID NOT CAUSE THE CLIMATE TO CHANGE? 86
  - ON TO A SMALL SHOW OF FACTS. LETS GET IT STARTED! 87
  - CLIMATE CHANGE IS NOT WHAT YOU BELIEVE. NO WHERE NEAR IT. 87
  - I HAVE SOME EXPLAINING TO DO 91
  - CLIMATE CHANGE STARTED LONG BEFORE THE INDUSTRIAL REVOLUTION. 94

TIME TO SHAKE THE TREE TO SEE WHAT FALLS OUT. WE GET TWO RENEWABLES BUT IN NO WAY EQUAL. 99

FOLLOWING UP WITH CAUSE AND EFFECT 101

AN ESSENTIAL COMPONENT OF THE STUDY, GLOBAL TREE, AND FOREST AGE. 109
IT IS WORSE THAN YOUR THINKING. 117
A PEEK AT SOME RESULTS ANYONE? 119

DEEP DIVE WITH A WARM BRAIN 123

FIRST, AN OFF-RAMP BUILT FROM TIMBER 123
DATUM—WHERE IT ALL BEGINS TO REALLY START TO SUCK 124
CONSTRAINED DEFORESTATION 125
THE AMERICAS' WERE LOSING THE LAST OLD-GROWTH FORESTS TO DEMAND DRIVEN FORESTRY. 126
LUMBER'S NOMINAL MEASUREMENT SYSTEM, ARTIFICIAL DEMAND. 129
WHEN ENOUGH IS ENOUGH 133
THE CONFESSIONS IN PPM DATA 135
THE PLANETARY GROWTH CYCLE 139
CO_2 SINK CAPACITY IS ALSO ESSENTIAL TO CLIMATE CONDITIONS FORMING OR BEING ELIMINATED. 144
KEEP IN MIND VOLUME AND CAPACITY ARE SIMILAR BUT DIFFERENT. 145
LET'S TUNE-UP EMISSIONS, PERMANENTLY! 147
ESTIMATING EMISSIONS' EFFECT ON CLIMATE 148
SUMMARY OF CALCULATIONS, THE FINAL TUNE-UP OF EMISSIONS: 149
I FEEL OBLIGED TO TOUCH ON THE FOLLOWING FINDINGS, AGAIN UNAVOIDABLE. 154

MAKE BAD NEWS WORSE? YES, I'M GOING TO DO THAT EVEN THOUGH I DON'T WANT TO. 159

UNFORTUNATELY, THE STUDIES SEQUESTRATION UNDERSTANDINGS ALSO PROVE HUMANS ARE RUNNING OUT OF TIME MUCH FASTER THAN THE UNITED NATIONS CLIMATE PROJECT PREDICTS. I DID SAY MUCH FASTER I MEAN A LOT FASTER. 159

TODAY'S 2-DEGREE WARMER IS TOMORROW'S 10-DEGREE WARMER, AND IT WON'T STOP THERE. IT CAN'T. IT'S A RUNAWAY (WITHOUT CMS'S SEQUESTRATION CURBS).	162
A SIDE TRIP AS AN EXAMPLE TO A MAYBE	165
SCOTTY, I NEED WARP DRIVE NOW!	169
TIME FOR APPLICATION OF THE STUDY MADE PRACTICAL WITH THE DREADED MATH!	174
IT'S HAPPENED BADLY BEFORE, STOP HELPING IT NOW	**179**
THE DARK AGES ARRIVED IN ONE WEEK	180
CRASH! IT'S NOTHING, BUT I CAN FIX IT	**191**
SO FAR, NO GOOD.	191
LET US GET TO WORK IN A POSITIVE DIRECTION BECAUSE WE CAN!	199
MORE ON FORESTRY AND WAYS TO WIN	201
CMS ECONOMICS FOR THE FOREST	205
SOMETHING FOR EVERYONE	208
NOW, SOMETHING FOR EVERYONE	214
NOW, SOMETHING FOR EVERYONE FROM LITTLE OLE ME, PERSONALLY.	219
TIME TO CLOSE UP... FOR NOW	**225**
GLOSSARY	**231**
BIBLIOGRAPHY	**245**

> Words in *italics* appear in the books "GLOSSARY" at the end of the book. These include some new things to ponder as well as expansion of terms shared with other disciplines. I do hope you find them useful in definitions of the study's findings as they provide additional context that can be difficult to explain.

ACKNOWLEDGEMENTS:

Beverly E. Law, Tara W. Hudiburg, Logan T. Berner, Jeffrey J. Kent, Polly C. Buotte, and Mark E. Harmon. Without their work, and Dr. Law's public statements on forestry emissions, my work may not have been possible. Their paper, "Land use strategies to mitigate climate change in carbon dense temperate forests" is groundbreaking and highly inspirational for Complete Mitigation Science's ability to define and model the cause of climate change and its mitigation. Thank you for pointing out what I call the *"carbon hump"* I was truly in need of that, and it found me.

To my wife, you do deserve better, but you're stuck with me. We'll have to figure that out within another 30 years.

Dave Felstul, thanks for being a scientific sounding board, and a friend of the family with writing skills.

To everyone involved with the Engineered Wood Company, the elephant is the only thing on the menu. Thanks for supplying the BBQ sauce! Mike, Andy, B&B, Michele, Thomboy, Theo, Van, Sybil, Roger; to all a big thanks!

Isaiah Ray, a great friend and even better listener. Where is that green couch?

LET'S BEGIN WITH REVENGE

I had a reoccurring complaint when others evaluated this book. From that complaint arose the killing of two birds with one stone. An opportunity to make the book better and extract a little revenge; so, I've taken advantage of the opportunity.

The complaint was I didn't dive right in and explain the science and impact first and foremost, and an editor said that I buried the lead. Okay, so I get it. I also read the summary first and then go after the details. As a fix, I've placed a funny but accurate summary on the first pages and guess what? No more complaints!

The revenge, you can read about in the summary; but, to foreshadow I'll add this: The below is a "fictional" summary completed with "nonfictional" materials. Why I say it like that is more accurate than saying the actual writer's ego was damaged by CMS. So, the writer, and as agreed, wishes to remain anonymous to avoid further scientific embarrassment. Thanks, anonymous, and as also agreed, CMS forgives you.

THE END OF CLIMATE CHANGE?

My colleague asked me to complete a summary of his work. Why he asked me was plain enough, revenge. I also know he asked others for the same reason. Cumulatively, I was put forth as sacrifice to his study, I lost the rock, paper, scissors, and Spock tourney.

I had been introduced to his work years ago but without any details. At that time, it appeared he had grown intent on committing career suicide. At least that's what I believed. From what I could get out of him and see for myself, it really appeared to be the unnecessary punishment of a long dead horse. Granted, I was looking from a distance but it sure looked like a whip in his hand and smelled putrid. His self-proclaiming of a "masterpiece" led me to constructing my "somewhat" disapproving judgement so it was all his fault. I was justified. At least I certainly was at the time.

Over many years I watched and heard of his struggles from mutual friends. And I laughed with them as I heard of his misfortunes. I'd even occasionally, and reluctantly, conversed on modeling methods and other tools when he'd sought my experience. I was really humoring him because the number of data points really limited his focus and he sought a way around. He needed a supercomputer, not me. So, my continued laughter and innuendos were not offers of help. And unfortunately, I only viewed bits and pieces of what he was working with. Which gave me ammunition to shake him down at every opportunity. I really poked at him with a stick sharpened by my and other "friends" wit. But I may have come up with that nick name that wasn't meant to flatter. I can honestly say I did all that because I cared. Not at all. If I cared I would have helped more.

And then he went and began the publishing effort and ruined all the fun. I vilely remember thinking he had a lot of nerve to ask if I'd help edit. I said no way, followed by some derogatory comment while using the nickname I'd assigned him; but the SOB sent me a manuscript anyway. And laughed when he told me he would and then hung up on me. Blatant disregard to my answer! But not like him at all, something was up.

Thinking I needed a laugh, I did glance through it a couple of weeks later. And then I read it repeatedly. It nailed me to a cross and left me to the crows. All of a sudden laughing at him became regret. Still, a little funny, maybe? Not really, not after seeing the entirety of the study. It was unexpected, new, and significant. And not a useless flogging of rotted remains. Instead, it was the BBQing of an elephant to feed our overpopulated village. Damn, I just realized, I am the idiot he describes in this book. Aren't I? Or at least one of them.

After a month or two of my reluctant review I called him and hung up on him when he answered, three times. I only wanted to leave voicemail and certainly nothing written that admitted my defeat. He received my message, I refused to take the returning call.

Sometime later but not near long enough. I was tricked into answering a call. He asked if I'd do a summary for his book, at the time untitled, this book. He wanted me to write "a forward" and I again said no way followed by, "don't rub it in, a-hole." He obviously had a plan because he suggested it could serve as punishment for my years of deleterious behavior and suggested, "all would be forgiven." I replied, "you do not know what deleterious means...Fine! Give me a month or so but it will be by anonymous." and I hung-up. It did not take him long to guilt me into writing this summary. Really, I might have done it if he paid for lunch. So here it is, and all is forgiven, "you tree loving nut job."

For the record, Tim edited this for accuracy and helped by urging a point or two missed. There are so many points, it's still unbelievable how they all lined-up to disentangle and reveal CMS. I also suspect he will make me out as poorer quality than I am, since his edit is last before this goes to print. Maybe mercy will prevail. Tim-only if you had signed it!

Cure climate change now that we know what it is? Really? I mean is that possible? I could not believe that even after I was years ago made privy to parts of the study. But yes, that answer evolved from no way no how to a simple yes! But that answer also comes with a "who knows" and a couple of "maybe's" firmly attached. The who knows and maybe's arise because of new knowledge that revolutionizes everything climate change related. Which all boils into a singular point of contention, "climate change is not what we all believed it to be, its worse. Much worse."

Let's start with the results of the study by explaining what makes it notable. The study, "Complete Mitigation of Atmospheric CO_2 From Residence Conditions and Emissions," is its formal name, "Complete Mitigation Science or CMS," is how my colleague and tree loving nut job, Timothy C Thompson, refers to it. Okay, he really isn't "a tree loving nut job." He is just a guy who figured something out before anyone else did, at least that's how he modestly refers to himself. Let me put it this way, everybody used to call him "Tim." But now we all call him Mr. Thompson, and he hates it. But we do not!

Anyway, the "CMS" study turned up unknown and interesting facts about climate change. What makes them interesting to me is how they surfaced from data not usually associated with climate studies. The unusual data source is from historical forestry uses and their correlation with atmospheric CO_2 parts per million (ppm). These unusual sources of data become highly relevant by neatly stacking logic into the study's cause of climate change as an environmental impact.

It seems that many widescale deforestation acts are documented in history. These acts were first recorded in stories and historic accounts from the likes of Gilgamesh, Enkidu, and others from the Mesopotamia era. Mentions of Phoenician's building ships that put an end to Lebanon's great cedars is just one. The Minoan civilization collapsed due to the desolation of its forested lands that stopped their fuel for copper smelting another of many. And many, many other accounts of human domestications forestry demand like those are found globally. Today, it turns out a good thing that our predecessors recorded their forest abuses ever since writing replaced cave art. The study provides other examples as well.

The Roman's understood the value of wood as defined by Pliny's books XII and XVI of "Natural History" both devoted entirely to trees but with lesson's learned too late. They were written after the complete stripping of Italy's forests. The Roman Empire became an Empire because it required charcoal from the forests for metallurgy and those economics added into their goal of conquering "wooded lands." In China, signs of wood shortages appeared in the early 13th century and continue today. The Spanish Armada made the study's point when in the 1580's its construction left sizable portions of Spain and neighboring countries devoid of trees. Britian and France also began importing trees with most of Europe by 1620 to build ships and heat their stoves. By the early 1700's mature trees were all but gone so Europe entered into an energy crisis. By then, luck was on the human's side with the America's becoming reachable with larger and better wooden ship technology. That's something China had possessed century's before Europe but building them had devastated the Pacific's forests so they could not be rebuilt after their useful lifecycles.

1600- 1700 is when the America's began opening up and supplying European forestry demand. The demand for timber and wooden ship building skyrocketed and so did the high demand for products made of wood. Forest use intensity did not stop there either. It continued to grow with steam power fueled from the America's "inexhaustible" forest resources. At least, that's what the people at that time believed. But railroads had not been invented.

The America's and the globes railroad expansions that began in the 1830's put an end to that "inexhaustible" thinking as railroads consumed all remaining of what the study coined as "convenient forestry." By the turn of the century human's again were quickly running out of easy to get to forests. By the late 1800's humans began tapping the last mature forests on the west coasts of the Americas to supply its growth rates, they had already used-up the east coast forests. Today forestry demand is fueled with modern items like furniture, fast food packaging, paper plates, cups, drink holders, houses, and toilet paper. It is also filled with artificial demand from "nominal measurement" and the current "Mass Timber program" and other "believed" renewable resource programs like Bio Char and Cogen plants. The study makes it point as both those lists combine with increased human population, ongoing forest immaturity, and changing land uses to highly restrict, impede, and decrease terrestrial CO_2 sequestration. Buy millions of percentiles. And that turns out to be a dangerous kind of thinking.

Why? Because none of those forests and billions of other forest acres have yet to recover into what would be considered "normal mature forestry." In short, they have been rotationally harvested ever since man cut down the first tree within them. Many forests have been harvested hundreds of times and each time they were, they lacked maturity.

The study candidly points out that CO_2 and $CH4$ driven climate change actually began 7,000 years ago with demand driven forestry practices engineering it. Even more to the point, the study demonstrates climate change as an environmental impact. A result of geoengineering. And wait for it...Not the CO_2 emission-based problem we all believed. Instead, it explains, "that climate change is more "sequestration dependent" than "emissions dependent."" Which is a vastly different way of explaining climate change. But here's the thing, the study's findings are exceedingly difficult to argue against. They do check out and in more than one way.

To start, the study defines historic forestry use with its coined term "demand driven forestry," those practices that substituted forest "maturity" with younger and younger "immature" forests. Today, 85-95% of global forests are under "demand driven forestry," practices. That very human impact prevents forests from maturing and was determined by the study while documenting average global tree ages. What the study found is average global forest age is typically something much less than 45-years-old. And those immature forests are highly subject to harvesting at any given time starting around age 28. The study makes that point sharper by directing attention to a little-known fact. The CO_2 sequestration ability of a forest or a tree is proportional to its maturity in years and its mass. The study opened a conversation regarding the lack of tree maturity effects on climate change. And since CO_2 levels in the atmosphere is the climate changing condition the study proves its point again, again, in more than one way.

    The example supplied is a typical 30-year-old tree can sequester around 150 lbs. of CO_2 annually, while that same tree, 70-years later, can sequester 1400-1800 lbs. annually. The tree increases its mass/sequestration volume at around 3-8% during its annual growth periods. Results have mitigating factors like tree species and geographical location but the typical forest specimen growing today fits the bill. And as if that were not enough to think about...

Continuing more than one way, and something to consider seriously is the above concept globally expanded to check the study's work. It is a revealing statement about the study's results. "If you could remove humans from forestry, historically, but kept all the emissions humans have produced, also historically, there would be no CO_2 driven climate change today." Forrest maturity would be opposite of todays, so sequestration ability would exceed all emissions levels. So, CO_2 driven climate change within that scenario is not possible. A very weighty analysis of climate change to say the least. The study does not stop there either.

The study says that scenario works because human emissions are less than 1 percent of the actual climate changing condition, which is actually a human-caused sequestration shortage. To demonstrate, the study points out two CO_2 emission factors and an atmospheric condition to consider. First, natural CO_2 emissions are estimated to be between 400-750 giga-tonnes annually. Second, human emissions are really only 10% or even less of natural emissions levels and estimated at 35 giga-tonnes yearly. There is one more key factor to consider as well. Over 3,000 giga-tonnes of CO_2 are currently within atmospheric residence conditions. That residency is keeping Earth's biome well above a desired CO_2 ppm level. The desired level being in the low 200 ppm range. Today, Earth is at 430 ppm. So, something like 1,400 giga-tonnes of CO_2 need to be sequestered from Earth's atmosphere to mitigate climate change.

The points the study revealed about emissions are made valid in ways not anticipated: First, humans can do nothing to reduce natural CO_2 emissions, they are in fact, untouchable, and as the song goes, "that's life." Second: realistically, and in fifty years humans have not been able to do anything about their own emissions. The study actually extends reasons for those failures. It goes on to explain humans as "emission dependent" and human domestications requirement of emissions. So, emissions are necessary, needed, and are therefore 100% unavoidable. It says, "everything human is CO_2 emitting, even breathing, so emissions are, for the most part, unavoidable."

Another sharpened point of the study is emission reduction plans do nothing to address current atmospheric ppm levels and really only leak CO_2 themselves. Since CO_2 is the "what" in "what" causes the climate to change. Residence conditions must be addressed in mitigation.

Now, on to the study's insightful punchline; sequestration efforts addresses 100% of CO_2 sources and atmospheric residence conditions. It does so quickly, globally, and economically. And its proven, not theoretical like CO_2 emission reductions are (for a climate cure).

According to the study, Earth's biome's emission sources and atmospheric levels turn out to be variables on one side of the climate equation. Being so, the other side is made obvious and more important, sequestration. When released, CO_2 has to have someplace to go or else. So, CO_2 will accumulate in residence conditions if not sequestered. In contrast, "even if human emissions stopped today, natural emission volumes and the heaps of CO_2 in residence conditions would still be accelerating climate changing conditions tomorrow.

The study adds to that analogy with two important reminders: Within Earth's biome, and by a law of conservation, Earth's climate change occurs conditionally and within a closed system. It therefore makes perfect sense the two sides of the equation must balance for homeostatic conditions, or balancing, to form. Which also physically explains sequestration's current deficit adding to today's atmospheric levels. That also marks why sequestration must be addressed in mitigation to actually end CO_2 driven climate change. As the study overwhelmingly puts it, "there really is no other way to influence CO_2 driven climate change but sequestration."

The study really did "isolate" the cure. "Mr. Thompson" calls it "solving the climate change puzzle." But how does it rate on the historic science announcement level? As a kind of an admission of guilt but relevant to my rehabilitation; my take is, there are several parts of the study that stand out and deserve recognition for their revealing. But when you combine them all together it unlocks climate changes secrets by creating new ways to think. I predict, time will be in the study's favor but it will take time before it is. Its fact driven and not emotional about anything, that is to its greatness a disadvantage.

Continuing on with more than one way and micro analysis of another empirical. Another subject of the study's is a gentle reminder about photosynthesis's use of CO_2. Plant life evolved to be far more accustomed to shortages of atmospheric CO_2 than too much. As explained, not enough CO_2 is the norm for photosynthetic production; However, too much creates plant growth problems with excessive CO_2 fertilization.

That previous point is empirical by reference, practice, and measure. The study can therefore state, "Earth has definitely shown plant CO_2 sequestration levels evolved to exceed emissions levels." The study also entails, "plants evolved their not enough CO_2 attitude because of "normal" atmospheric residence conditions restricting both time and duration that CO_2 is made available." So, to plants, CO_2 is a hit and a mis and otherwise a completely random act of the plant being able to absorb CO_2 from the atmosphere.

The study goes on to say that is not what's happening today. Today, the sign of excessive CO_2 fertilization globally should be detectable in tree growth rings. And that is one part of the study the researcher wishes to expand on. He theorizes another correlation will solidify because excessive CO_2 fertilization lessens the amount of H_2O the plant uses. So, he has a tell to look for. I think he'll find that one as well.

Here's another of the study's sharpened points, sequestration didn't use to be in such a deficit until human domestication and populations took off. For hundreds of millions of years sequestration ability exceeded emission levels naturally. This is also where the study linked in the mother of all climate change problems. The study coined this problem as "impeded fast cycle CO_2 sinks." These are the terrestrial CO_2 sinks that should naturally exist in forestry and trees when mature enough. They do not do their job well, if at all, when they're not mature.

The "impeded fast cycle CO_2 sink" is an impeded fast cycle sink for a couple of reasons. First, trees lack maturity which decreases the sinks sequestration ability. Second, trees are not allowed to mature under current "forestry demand driven practices." What gauges them as impeded comes from comparing them today to if human's had never harvested that tree or forest. Which can be millions of percents impeded when approached either historically or calculated by tree life spans.

The study isolates impeded sinks into categories:

One, lack of sinks due to 10,000 years of changing land uses. 38% of the global forest has disappeared. The majority of which happened within the last 200 years.

Two, their shrinking capacities due to the ridiculous number of immature trees stuck in what the study coined as the "carbon hump." Prior to explaining the hump, there are trillions of trees on Earth. Estimates vary between two and three trillion that occupy an estimated 10.3-10.8 billion acres globally. Unfortunately, a vast majority of those trees are under the age of 45 and extremely immature to their species life cycle. Fortunately, around 1.1-1.3 billion acres does contain maturity as old growth, hopefully. All that maturity is in the form of National Parks and protected areas. In reality, half of that figure is still subject to logging and land modification, it's located within the Amazonian Forests. Much of the other half is only recently protected, so highly immature but growing in the right direction with no guarantees. All of which is not protected as much as it deserves. And the study sharpens that detail by saying only 3 to 5% of those lands actually contain mature forests. However, those remaining mature forests' sequester a minimum and estimated 92-98 giga-tonnes of CO_2 annually. Which is not near enough to slow or steady the 435-780 giga-tonnes of annual CO_2 emissions, So we need more trees just like the mature forests contain.

We only have those lands because around 1830 CE the inferior fuel coal and its improved version of coke began replacing difficult to find but the much better fuel, wood. Coal eventually became more economic and became the only known historical event that lessened forestry demand, thus far. Thankfully, humans didn't strip the Earth's forestry back then. They sure could have. Fact is, if humans didn't have that 3-5% of old growth doing such a good job now there is little doubt humans would have achieved a climate driven demise then. But coal really only postponed it a century or two because forestry demand has only increased in line with human population. We are worse off now than before coal's global introduction when it comes to sequestration.

Three, the existence as an unhealthy sink because of past reoccurring harvests events and the coined term of "tree degradation." That's explained by the study expanding land degradations more common definition. In the study, a 300 ft tree grew and continues to grow on unharvested land. The same land but regularly harvested can only produce a smaller 200-foot tree or none at all. Nutrients degraded, soil compaction, and the continued harvesting promotes climate change and causes drought. The same decades of droughts the globe has experienced and are recorded as today's events.

The study coined another definition that environmentally describes impeding forestry sinks as an impact of "constrained deforestation." This term demonstrates the logic required to apply the study to real world mitigation, just inverse it. "The action is thus constrained by the destructive circumstances perpetuating its results." The first sin pointed out is substituting a huge mature tree with a smaller immature one. The insanity conveyed is expecting the smaller tree to perform just as well. For the study's record, "one limb of a matured tree can sequester more CO_2 annually then 4 or 5 immature trees of the same species." Furthermore, one old growth tree can sequester more CO_2 than acres of immature trees or tens of square miles of replanted trees.

The "constrained deforestation" effect is constrained by humans using only immature trees to supply "demand driven forestry." And by not using forestry efficiently. Ignoring maturity again ignores both mass, as a tree size needed to process biomass efficiently, and sequestration, to address what the study coined to describe both renewable resources acquired from trees. The study coined a description of the sequestration resource as a "binary restricted resource." Its binary and restricted because the tree's biomass regenerates much faster than its other more valuable and also renewable resource, sequestration. That is also influenced by an unsympathetic and unfavorable problem known as the "Carbon Hump."

The "Carbon Hump" adds restrictions to both impeded and binary effects within forest recovery timelines. First, on average, less than 20% of biomass from a typical harvested forest keeps the sequestered carbon within products produced. At best, those forest products that used that 20%, sequester carbon for around 40 years. And because of the energy needed to create those products they are questionable at being carbon neutral and more often than not, they are carbon positive for their entire lifecycle. The remaining 80% of biomass harvested is undoubtedly carbon positive, an extremely high CO_2 emission source. The basic understanding given is that the harvest site mess and products produced from that 80% of the biomass are immediate CO_2 and NH_4

sources. Manufacturing processes, energy consumption, and global distribution also add CO_2 to both emission ledgers.

The study points to a given fact on forest products. Even after replanting a harvested forest plot (the majority of which are not replanted) the emissions from rotting biomass like stumps, roots, tops, and limbs plus energy consumed release more CO_2 and NH_4 than a replanted tree's (on that site) can sequester for 20-30 years. The man made "Carbon Hump" cannot be overlooked . It's a barrier to cross and the reason why forest maturity is the only legitimate CO_2 negative architect that can make new carbon sinks, or fix existing sinks. What the study also highlighted is the humps effect on climate mitigation. The point of logic is so sharp, its jab goes deep and really hurts.

The study found a thirty-year-old tree that's over the hump is really, really, hard to find even globally. Now, wait for it because the study goes deeper than you might have thought. It shows how little of the global forest is actually sequestering CO_2 and is acting more as an emitter. 85-95% of global forest is actually a CO_2 and NH_4 industrial sized emission source due to immaturity, positive emissions from harvested plots, and countless unplanted acres in natural regeneration (allowing nature to reseed from adjacent immature trees). The study's findings indicate the majority of global forest is within the hump's long-term grip. What I got out of all that is global forests lack significant maturity to be an effective CO_2 sink. After the study burned that into me, it took it up a notch and explained why climate change is worse than I thought.

Being worse comes down to applying the study's findings, or new knowledge, to old methods. The study's sequestration side of the equation is most definitely new knowledge to apply. I can see how applying a sequestration side to emission-based or global temperature estimates influence their predictions. In example, interaction with sequestration's current deficit and the study's datum and timelines. Doing so intensify earlier climate estimates and ring climate changing alarms louder. The study says it does by a factor of ten to fifteen times. It explains the effect of doing so. "Climate change has become the nightmare that science said it would, and we may have found that out too late." Answering the "what if's and could's," is how the study justifies that secular statement.

Remember where we started with "The who knows and maybe's arise because of new knowledge that revolutionizes everything climate change related." The "who knows" refers to if we have enough time left for CMS's mitigation plans to make the difference needed. It also implies that CMS mitigation will become palatable for bureaucrats and politicians but only after it's too late. The maybe is sourced from a couple of the study's facts.

Surprisingly, this book on the study brings up subject matter that touches on the challenges the study has hit outside of the science involved. These challenges are real and he describes them well. Which is a writing style I find unusual for a nonfictional and scienced based book. However, the book does good in explaining how trials and tribulations of any good idea and the best and brightest who are not in charge affect the study. He doesn't give names. I'm pretty sure he makes an example out of yours truly. I admit it plays out and is somewhat entertaining but sometimes an "off topic" attempt at drama. Which I think is reason enough to ask for this opening summary. I suppose keeping the book on point is important.

All the environmental drama generated by the study's use of sequestration comes from within its measurement and modeling of atmospheric CO_2 ppm's. Which also sends up the same type of S.O.S. rockets used by the Titanic. This section of the study drove a wooden stake through my toilet paper, log home, and new furniture loving heart. It became dark and that happened really fast. What first ticked me off was the lack of emotion it presented me with. The study made its point with such ease it is easily accepted, then concerning. Plus, it lays them out one after another, so again, in more than one way. And it also applies a maybe.

The study nor the book produce a desired result when adding sequestration into the United Nation's estimated global temperature increases. That turns out to be about half the time and more than double the temperature increase the U.N. has projected. In short, the study demonstrates hitting a point of no return from achieving an extinction level event by incorporating sequestration into U.N. data. It shows Earth entered a runaway greenhouse effect in the 1950's and began a CO_2 driven climate collapse around 1850. The book tries to explain those tidbits empathetically but it comes off unsympathetic, I for one understand why.

Today, naturally occurring volcanic events like 436 CE's mini-ice age, and the year without a summer in 1816 CE have become way more concerning to any reasonable and intelligent person. Why? Simply because we have almost double the CO_2 ppm today, a runaway greenhouse effect, a climate collapse, and little chance for recovery when those events happen again. And the book is quick to point out they will.

The study details that our climate woes are all because humans incorrectly believed trees are an easily renewed resource, as a renewable resource. According to this book, it explains the study's point as that truly turned out to be half of a lie and a trap because of the "binary restricted resource" definition. And Mother Nature is hell bent on our extinction; so, she baited that trap with an incorrect "renewable" definition of trees. One easy to believe and so very wrong. We took it, and now she's got us right where she wanted us, struggling, while she preheats the oven and cackles with the jovial satisfaction of a witch.

Back to more than one way. The correlation between that witches trap and the other points of logic explaining climate improve when surrounded by the atmospheric ppm levels the study analyzed. That makes the study not easily denied, really, not at all. Unless you believe the air bubbles in ice core samples don't contain a chronicle of antiquities atmospheric conditions. Since they do, they were used by the study to make its multiple historic points and develop its logic. But it also used Mauna Loa's more recent atmospheric observations (CO_2 measurements) from the 1950's to today. Again, more than one way.

The study's atmospheric ppm analysis statistically provides a serious challenge towards any attempt to debunk any part of the study. Believe me, I tried to upset the study. In more than one way. Now, I'm writing this summary as penance for having an especially uninformed opinion.

Even climate change nay sayers can't deny it. Not without lying to earn a well-deserved embarrassment later. The study is helping anyone who bothers to understand climate change. It does in a way that is very relatable, even to the nay sayers. It explains the unexplainable and easily denied emission-based science. The study explains that effect by saying, "sequestration proofs easily replace CO_2 emission theories." What I got out of that is there's zero speculation or opinion, just facts. I've also explained that as "no wool to pull over anyone's eyes is presented." That to me, and by itself, makes the study a leap forward in climate change by applying the cure to the cause and not chasing a symptom.

The empirical ppm data from NOAA is analyzed further and carried out by focusing on atmospheric CO_2 in and outflows, to test a theory. NOAA's Mauna Loa data was compiled from 1975-2021 to lock the test down. Doing so created compelling evidence of failing sequestrations existence. And in keeping with this articles scope, in more than one way, allow me to introduce how Mr. Thompson measured something that was not so easy to measure.

My quick summary of the theory to test is "During Earth's plant growth cycles, known as the fast cycle CO_2 sink cycle, the sequestration rates should be almost steady state when measured as an outflow from global atmospheric residence conditions and from one year to the next year." That was the thinking of science at the time. Annually, atmospheric CO_2 outflows should be in an almost steady state because of widespread plant growth on Earth. Essentially, we all believed the amount of CO_2 plants on Earth used every year would be about the same. I'm not going to get into why we thought that other than to say we shouldn't have.

What the study did to test it? Its own computation of NOAA's atmospheric data, the study went where nobody else dared. However, its results failed at establishing a trend line that agreed with the theory. The trend line that did reveal, unintentionally, was declining outflows that did not correlate to the increasing CO_2 emission trend either. That new trend appearing was unnatural to Mr. Thompson and sent him looking for why and how it was happening. It was for certain that something was affecting sequestration in some unknown way but he lacked a way to describe it for a couple more years. But he did, eventually, CMS.

The test, the "CO₂ ppm Delta" was the first part of the study that captured me when skimming that unsolicited and unwanted manuscript that needed editing. Mr. Thompson had somehow found a bolt in the heavens and proceeded to whack my cranium with such force that I'll never be the same again. Not when it comes to climate change. That lightning bolt made me stop skimming and take notice. Seriously, what else besides sequestration could cause that declining trend? Nothing, that's what. Okay, Santa Clause's magic dust, a sprinkle of which allows reindeer to fly. Yeah, that "might" do it.

To work from an unintended result is common in science studies. That particular one further helped the study develop its "CO₂ ppm Deltas" models. Which in time, provided the way to measure and rate global sequestration quickly and accurately. An intended result that just so happens to work well enough to gauge sequestration's global health.

Of all the study's components, which have appeared to date, the ability to establish a year-to-year sequestration-based trend is significant. That trend revealed the accelerated decline of global sequestration.

Now the bad news. A sequestration rating can now be determined with the "CO₂ ppm delta's." It's bad because those ratings went negative in 2016 and again in 2019. That is a very globally boiled over reality appearing for the first time and concerning enough to have induced a panic within myself.

Typically, the "CO_2 ppm Delta" would always be positive but closer to today it appears tapering to small or meager digits compared to pre 1800 CE amplitudes. So, those negatives appearing recently also indicate almost "zero" sequestration took place. Almost zero between the high inflows and little outflow that happens before or during the ppm measurement occurred. So "some" sequestration did occur, from the 3-5% of old growth and ocean. However, the norm is being measured at a global scale between peak inflow and peak outflow, creating a delta (the change between the two points). Hence negatives occurring beg to be concerning because they indicate a disparity, a linear crash of the outflow peak establishing itself less than the inflow peak. Another way to say this is the delta negates the constant and ongoing in and outflows and establishes a "norm" measured only during the growth season. By doing so, it implies another reason why mitigation, although possible, time is running out, "maybe."

THAT event the study recorded twice had never happened before. Not while humans have existed. Likely, not anytime in the previous two hundred million years of balance. The study theorizes the times that negative sequestration happened before was just prior to any one of Earth's six previous extinction level events. At least the events involving CO_2 concentrations. So, the study knows Earth is well practiced at that sort of thing, extinction being that thing.

Those negatives also tell us Mother Nature has warmed up to do her part. That role is of course, killing us off. But she's a lazy old hag; she's counting on us continuing to do ourselves in. She also has allies who are not so lazy. Time and human nature are for sure on her side and surprisingly bugs are another of her dependable allies.

A biological effect of the climate changing allows tree killing beetles to expand their range into higher elevations. Normally, those elevations are too cold. Nonetheless, over the last couple of decades the beetles have entrenched themselves into new goldilocks zones. These bugs are now wiping entire forests off the map and accomplishing the destruction overnight. In addition, the dead and dying tree's the bugs leave behind are perfected match sticks intent on striking up forest fires. And because the tree no longer shades the forest floor highly flammable and densely packed under growth swells and provides the perfect kindling.

They await a careless human or lightning strike to turn higher adjacent elevations and bug free forests into ash.

Forest fires have become more of hell's sought out reality than ever before. Not because there is more of them and they can be immense in size, it's because they too are an ever-expanding and spiraling consequence of climate change.

Forest fire size and ferocity is about how much moisture the forest holds. Well, immature trees burn like roman candles on the fourth of July with their highly flammable and compact limbs close to or even touching the undergrowth and each other. Plus, they hold extraordinarily little moisture compared to mature trees so they dry out quicker in higher temperatures and drought. In contrast, mature trees' limbs don't normally touch the undergrowth while their thick bark is typically fire resistant or even fireproof enough to allow undergrowth to burn without causing it damage. Some mature trees even need an undergrowth fire to help release their seeds and establish new growth. Finally, mature trees naturally space themselves with decades of natural selection, you don't lose acres if one unfortunately burns to the number one cause of forest fires, lightning.

The decline of forestry that makes it not forestry now or later is coined by the study as "unconstrained deforestation." All those climate impacts are defined by the study as increasing unintended climate demise in an ever-expanding spiral.

In summary, the more "unconstrained deforestation" occurs, the more climate degradation, which leads to more biological biome changes, forest fires, adverse weather conditions, and more climate change. All of which in turn results in further "unconstrained deforestation." The action is unconstrained in creating the destructive circumstance and perpetuating its result. That result is permanent deforestation. And yet another reason why the study can only answer honestly with a maybe. Time is not on our side.

To me, the "Complete Mitigation Science Study," is the best definition of climate change conditions put forth. It's dictation of environmental impact seems reversible, hope filled and needed points. And yet, the messages it conveys are just as terrifying when reviewing the damage from an F5 tornado that won't dissipate back into the clouds and stay where it belongs.

I think the problematic side of this study's message is that it just makes too much sense, which in a way, makes it unbelievable when first hearing it. But the thing about standing for facts and logic in messaging is that they can and do take time to be accepted. Unfortunately, I made myself the poster child by not realizing that early enough. CMS's proved to take time with me, a couple of weeks after my exposure passed before I sat up in bed and said, "that SOB, really did do it."

When writing this summary for "Mr. Thompson," I couldn't help speculating on the study's future. Humans now have a much better understanding of climate change and I don't say that lightly either. After 50 plus years of chasing a cure, it now seems apparent it was there all the time but hidden nefariously by Mother Nature, as the study puts it. I say it's that looking dumb thing that occurs when something simple is pointed out and embarrasses us.

The proofs that argue the study's points make the logic of it all almost surrealistic when learning of them. What I'm saying is I "feel" as if I could have easily come up with what the CMS study can now broadcast. It's only the realization that I didn't and nobody else did either can maybe be explained by the study's multiple variables being difficult to fit together. And what? Eleven new or expanded definitions that explain it all. Is it really this simple? Still a very eccentric or even unconventional but occurring thought with me. An embarrassing thought, totally.

I had to accept that without the terminology defining its logic we were all looking the wrong direction when it came to climate change. Admittingly, "Mr. Thompson" made me look, sound, and act stupid until he began pursuit of publishing; and now I'm totally hurt, and angry.

Allow me to dry my tears and be an adult. Months after seeing CMS but thinking about it all the damn time, there is something the study is that Mr. Thompson would never say. It's greatness in acts and highly accomplished by its results, there I said it, it's over.

Now, how to live with my shattered ego while my beaten, bruised, and no longer valid climate beliefs still swim in my head, I'm not certain I know. I suppose I should start with a congrats to "Mr. Thompson." You really did do a "thing," as you put it. And sincerely from me; indeed, an exceptionally good thing for all of our sorry butts, you tree loving nut job.

Anonymous

PEOPLE PERSON?
NO, AND IT'S NOT YOUR FAULT

I'm an engineer/scientist, so I have a communication issue. I only communicate when good reasons outweigh my "inability to recognize the emotional needs of others." I really don't wish to ever offend anyone. And it's not that I don't recognize the needs of others and can't adjust my communication style. I can and do, I really do, but only after it's too late. All my life, I've been empathetic to a fault but always regretting the timing of my occasionally unfiltered responses.

I unintentionally make some exposed to my brainy personality feel inferior. My words don't suffer fools or the incompetent pretending to be competent. I only care for the facts. I also freely admit I don't have a monopoly on all the clever ideas or understand the logic involved in all good arguments. But an uneducated opinion just doesn't come from me, ever. I pride myself on that quality because if I don't know something, I won't just make it up as I go along. I prefer to express that I don't know, or I can find out. I learned to do that as I matured. I did so because of my presence in public and private meetings where my profession and I "are expected" to "know it all." I don't know it all, nobody does. Unless you studied it, you can't. And this book is based on a study report; so, I'm going to just be me and provide the facts the best I can. But me has been a bit of a problem to some.

I'm the guy who studies the topic before the meeting. I'm also ogre-sized, literally. So, my actions are of academic accountability that nobody wants for themselves. And my physical appearance scares children at first sight. I'm a big, mean, ugly, and confident ogre. Something children, adolescents, and adults don't wish to ever physically appear as themselves. I'm no beauty. A friendly demeanor? Not at all on the surface: I bark and have been in trouble for biting idiots after giving them off-ramps they refused to take. I don't suffer fools well at all, ever. Now, I can't recognize or understand how all that mixes to make me who I am, but I don't argue with the results. I know my current faults well and I am certain I'm not without them within the opinion of others. They are my own and have proven to be lonely, and at other times, highly desired. A calamity of a persona, yes, I know.

When suggested, or ordered nicely, I should author a book, this book, and explain Complete Mitigation Science... I could recite without much thought reasons to say, "Not only no, but heck no!" But my responsibility to CMS makes this book an absolute requirement and the others surrounding my work are too busy just surviving. OR their writing is worse than mine. Plus, credible nonfiction requires a heck of a lot of dedication that is eased with experience, I learned that drafting this book. Now, I can only add to that by saying here I am, your reluctant author, promising to give this book his all and screw up your current perception of climate change for the better.

What makes that goal a challenge is that Complete Mitigation Science is contrarian and not yet mainstream, which further complicates the book's message and makes it challenging to convey. So, I'm left with providing all the facts until it becomes mainstream in society, which is already starting to happen, and much faster than we anticipated. I attribute it catching on to the profound logic and common-sense present in CMS's message and facts. I really don't know how it's gaining popularity, but I do understand why, it's the logic shining brightly and parting the clouds from years of diversion. In fact, you might even wonder why you didn't think of it. Which means I did my part well because that's the reaction I'm hoping for. And it isn't an easy concept to relate to others. Well, some of it is.

The problems and solutions I communicate are more than black or white, true, or false. Each has relationships to facts that correlate. The standard rule of perception doesn't help findings like these, either. Then there's the problem of me being the person doing the explaining. Communicating CMS to anyone the first time is a problem, and it's only made worse by my issue of communicating anything. To which, I reply "if CMS's message saving your butt offends you, then please be offended." See what I mean about being me?

The good news is that people who read this book's draft thought it was worth the sacrifice. But they all reported being "unintentionally" offended (I added the "un" to it). It's never on purpose and yet, there it is. Even with a good editor I still offended just by being true to facts. Darn it all! I'm mystified by that reaction.

So, I'm left to claim responsibility for my words and hope for a fair fight with the reader's current perception. Plus, I pray the reader still likes me enough to stick around after the torture of my fact driven writing method. Others have, but I might have paid them in lunch, coffee, or consulting fees. One was my mom. Helping CMS save humans from human actions doesn't sound like a bad trade-off if you ask me. You know, a save-the-world kind of thing only CMS can offer.

Okay, where do we start? I suppose at the beginning of all this. I don't want to spoil the story or the study by starting at the end. We'll get there soon enough. So lets start with the drama.

JUST ANOTHER CLIMATE MOUTHPIECE WITH A TREE

When I first started, I was working towards inventing some renewable climate-friendly forestry products. Stuff like sheet goods and even lumber substitutes for building. Those things are made from something I named Advanced Woody Composites, AWC for short. Doing this appealed to my closet tree hugger, and of course, I liked the sciency work involved. My wife calls it my obsession with "scientific" things and all the associated gadgets and gizmos that go with it.

To be sure, that time in 2017-19 that particular AWC workload had started to blossom. It led me into a vortex of carbon disposal calculations. By 2019, it sucked me into years of sifting through good and bad science and writing notes on previous notes, and then into manuscripts and patent applications. Eventually, I even put it all together in legible mathematical models, testing, and really big plans.

My motivation to invent AWC stuff was easy for me and my wife to understand it's that obsession thing. But, because no plan survives first contact with the enemy, the enemy being me, my AWC math models made studying climate change my new obsession. Why? I found something I couldn't understand. And that is all the bait needed, if you want to catch my obsession with scientific things.

In hindsight, something else stood out to me that is easily understood. Failure. And that failure was becoming increasingly noticeable with tornadoes in December, heatwaves, prolonged droughts, and all kinds of newly recorded and record-breaking global environmental mayhem. And yet, the drums of the failing climate cures and their conscripted army of followers still beat on while sounding off their catch phrases and profit schemes as the artic icesheets melted. And even today they continue their chants while others are still in complete denial and still unconvinced with the failing party's decrees of completing climate suicide. I soon stood in lockstep with the minority of the unconvinced; but I posed "zero" hints of any kind of climate change denial. There is ample proof of CO_2 climate change existing; but, how to address it was obviously not working.

50 years. 50 years of failures has made exactly "zero" progress in even making a dent in global climate change. If anything, those 50 years of failing to influence the problem has made it much worse.

The year before Covid, the drummers of the failures changed their climate cure pitch into really bad excuses. I sensed their change in tune as desperation and gained the understanding that perhaps we should try and learn from our mistakes, 50 years of them provided many opportunities. That thinking allowed my obsessive science nature to take over. All because I had already seen in AWC's what I did not understand.

Not surprisingly, this obsession progressed like any other, transitioning more into isolating facts from fiction so I could learn more. Before that time, I cared about climate change, but not more than my daily routine or perception allowed. I mean seriously, global warming was somebody else's problem, not mine. People a lot smarter than me were working on that problem, right? My understanding was it's a complex problem, but emissions reductions or the tens of other "things" would eventually fix it, right again, right? I really believed climate change didn't need my gray matter or me jumping out of the nice warm box I'd made for my career. It was being taken care of! Maybe not.

Unfortunately for me, I was born with the curse of having an inquiring mind and not so dumb. Therefore, I'm quite capable of making uncomplicated things extremely complicated.

You see, I can't just buy a TV from the local store. I must research TVs for months before buying. I've been looking at replacing a vehicle I own for three years and at a touch-activated faucet for six. It's a curse. But when I do something, I do it well. Idiot savant? Maybe a little. But all that research made me experienced. Experienced enough to eventually recognize something wasn't quite right in climate change's origin story or where it gets its powers. It had to have a chink in its armor as well, a kryptonite.

For the record, I'm not entirely at fault for my obsessions. Usually, I carefully select my next obsession with my wife and colleagues, and this study screamed for all my attention. They supported the study even when I could not really explain exactly what I was doing or looking for. Likely because they know me well. They could already see I had the scent of something fixing my stare and sent me forth into the unknown. So, you see, I get permission to visit the darkened side of things. Because they know my mental absence from them eventually returns, at least so far it has? Okay, I'm not the easiest to be around, I admit it.

The next part of AWC work was supposed to be easy, even for me, a guy who can complicate anything. But that mental attention trap had sprung. One minute I was working on simple carbon content calculations, then wham! Into the more complex climate change studies I went.

Like everything else I've done, my simple tasks became an elephant to eat without sharing. I was solo on a big project, yet again. I'm stupid like that. But I really could see something significant, although I could not explain it, except with my evolving theory, conjecture, and other guesswork, which meant no facts to back up anything real, yet. But I was sure of one thing: the math was correct and pointing to the second star on the right, to Neverland. And what scientist doesn't want to go there? But then again, it is an exceptionally long flight!

Months of unfulfilling research later, my still blank mind could only stare at what I now referred to as "it." I continued poking and prodding to figure out what was happening and intrigued me so much. Little by little, I focused the spotlight and revealed a little more but quickly recognized I knew truly little of "not much at all." Still, I clung to the hope of reaching some kind of eureka moment but had no idea what that might look like.

The problem too work. AWCs' carbon neutral impacts were much more significant than I had anticipated, but not in the way I'd expected, which was good news but still mystifying, nonetheless. The tracks laid by that good news were unexplored. The sad news was that those tracks were strewn across rocks with tall weeds growing through them and wouldn't be easy to follow. I could see where they led but not where they began. The end was also somewhere out there, guarded by tall, prickly weeds of climates failures. The time to get off the data-pulling tractor and walk through those weeds had arrived, and the trek toward defining it all began with the first "ouch" of their stinging thorns that shredded the belief in emission reduction science. But how so? And how in the heck was that even possible? Those thoughts had been nowhere in my thinking, but the study started to reveal them as valid; And then the real question to answer popped up, "could it really be so simple?" Yes, yes it could. And there was a Santa Clause, once.

 I'd assumed AWCs' impact would be fantastic, but what the study revealed was significantly better than AWCs could do, or really, anything else out there could to cure climate change. "It" was viewable in the data because of the raw material calculations I added on a whim to expand the accounting of carbon in wood. The study no longer only looked at the products; it could now observe the remaining forest the materials came from. The 66% of the trees AWC's had saved from clear cutting. Sometimes you get chicken, and sometimes you get feathers. The study provided all-you-can-eat Kentucky Fried Chicken served by Batman's butler to me and Wonder Woman on date night.

It was clear that even minuscule changes in the raw material supplies over a period of years greatly improved "sequestration" by adding to the carbon accumulation within a tree or a forest. So, the older the tree, the higher the tree's efficiency. And not just a little, almost exponentially higher. This is nothing new to forestry. But the way that carbon accumulation grows explained something else, possibly something new, "it." To explain, I have to go back a little in the study's timeline.

When adding CO_2's atomic mass unit (44 amu) the study's new sequestration calculations modeled the conversion of CO_2 into a carbon (12 amu) within one pine tree. Sequestration exploded upwards and grew yearly. Again, the study didn't invent or discover how to do that. It's empirical, a known quantity. What the study had done, at this point, used well-established, empirically measured methods. But, by doing so, the model no longer solely considered only carbon storage. It now looked at the raw material supply's conversion of CO_2 into the wood's carbon storage. That changed my entire outlook on climate change because it revealed precisely what caused climate change, at the time and still in theory.

"It" is difficult to explain because there is no smoking gun. What I mean is "it" is not made up of one thing that is effective, "it" is made up of several things pointing to one highly significant thing. And "it" was hidden in plain sight and for reasons I can't explain, undefined or overlooked by science.

Literally, everyone took "it" for granted, they didn't know about "it," or maybe "it" just wasn't significant in daily life enough to care about. I know through discussions most believe "it" is a fixed number or stagnant and not the variable sequestration really is. Which pointed to more path to follow.

My supersmart offspring and I had a few conversations about what the study turned up. My youngest kid, Theo, is not a science person but is chasing a career to become a professor at a university and on a fellowship, basically not so dumb of a millennial. Anyway, we devised a description for the science around "it" that goes like this, "It's all been said and done before, but the disciplines involved have never been combined into a single conclusion as meaningful." That conclusion is what is causing climate change and how it started in the first place. Which by and large pointed to where the study needed to go next.

The study's facts and their results were now staring me down, but perception crept in, and they became harder to believe. Being human, I'm no stranger to disbelief. I remember thinking over and over, how could this be right? It's way too simple! How in the heck has nobody else seen this? They probably had but had taken it as being empirical and not as the variable it truly is. And here we go again, back a little to move the study forward. Two steps forward, one step back sort of thing.

The equation for establishing climate change, as in, being able to establish climate-changing conditions with cause and effect, needed an entire sequestration side added to become relevant. The study had already determined the cure for climate change was not just about emissions inputs or reductions, the energy consumed, or CO_2 leakages. Those now appeared only as variables that really appear only on one side of the climate cause and cure equation. Without sequestration, as the tree absorbing CO_2 from atmosphere, set equal to those other variables, the math just doesn't work out. That fact is one of the peculiarities of the study and difficult to absorb at first exposure. So, let me try and explain in a separate way.

To describe that, I think of a basic chemical reaction: every element on one side of the equal sign must be accounted for on the other side. If they're not accounted for on one side to the other, the reaction is wrong, it's physically impossible. With that, I had realized everything I'd previously modeled about AWCs climate benefit wasn't wrong, it was however highly incomplete. Adding to it on that earlier whim made it all seamlessly work. To get the equation to work for climate mitigation the study established sequestration on one side of the equation in deficit and in comparison, to emissions and the other variables I mentioned on the other. That was the thing that got me thinking that sequestration is more important than anyone in the science community would have believed or advertised. Seeing the sequestration in deficit was needed to balance the two equation sides demonstrates something special about climate change. The study now solemnly showed that how we all looked at climate change violated the *law of conservation*. And that segues nicely into what the *law of conservation* is. It cannot be ignored, or its sharp toothed logic will bite you, every time.

There can never be any more or less of any element within a closed system. The system is Earth's biome; the element is carbon. The element C can be transformed into molecules like CO_2, but you don't get any more or any less C unless it's added or removed from the closed system, like a meteor adding to the biome from space or a rocket ship sent away from Earth, never to return, deducting from the biome.

Earth's climate change occurs within a closed biome. Consequently, climate change is a balance between molecule conversion like fossil fuel converted into CO_2 as emissions. OR CO_2 converted into biomass. SO, it must be balanced between storage and emissions. Biomass is the first stage of a burnable fossil fuel and in historic fact, wood has been used more for fuel than even oil or coal combined. Because oil and coal were biomass before millions of years of heat and pressure turned them into the fossil fuels of today there is more convertible carbon there than anywhere. Now, how do you think fossil fuels became so carbon enriched before their storage? That would be from photosynthesis millions of years ago, the same photosynthesis plants and trees use today to balance Earth's biome under the *law of conservation*. In simple observation, if the carbon from fossil fuels is released as CO_2, then the conversion of the CO_2 back into storage, as the biomass it originally came from, is also required.

THE STUDY ONLY IMPROVED FROM THERE.

AWC's showed highly improved carbon benefits with a new and improved model that expanded the study. The study didn't realize the "sequestration" effect or the details surrounding it yet other than to understand how sequestration worked tirelessly against climate change. But here's the thing. The methods others used to check AWC carbon benefits were their own but improved with the new balancing. In short, checking the study unknowingly verified the math and the theory on sequestration's roles. It was around then I knew the study had done a thing for sure!

But it wasn't time for "Eureka!" yet. A long road still awaited. Proving a theory in only one way doesn't always mean it's right, one must have multiple proofs with repeatable results to be certain. Plus, the study's models were working single years tied to single AWC products. However, I could see the longer periods like decades and centuries. When expanding the model again I did find a second proof, and even a third almost as quickly. Each of them used different methods and even different disciplines, which was weird. I do demonstrate the proof and the weird source later. But right then, I was certainly gaining confidence!

I knew it for certain now, so further down the rabbit hole I went. Colleagues who checked my results didn't understand what was created in the balance, even after I subtly pointed "it" out. To be fair, I wasn't entirely convinced yet either, so I never said, "hey, look, it's right there! That's amazing, right?!" Although, I really, really wanted too. In today's retrospect, time for colleagues and my gray matter to catch up did pass. Some still haven't replied to my years-ago emails, which is peculiar but explainable. For now, let's say it's a perception thing everyone introduced to the study experiences. And I mean everyone. It took all of them time to adjust. It isn't easy to overcome years of training, regardless of how beautiful a new and blooming solution is. Still, it was a disheartening experience, and I took another break with another paying job, this time for seven months before that sidetrack to pay Bill ended.

Unlike other job's I'd taken, this one actually provided some "free" time; during which, I'd worked up a new climate theory to test. With it, I began to define all items within "it." I was now basically going after the logic to explain "it." Again, I worked independently but verified my theory that forestry demand affected sequestration. But what I found out about climate change wasn't the problem. People were.

I expected some backlash to the study's findings but nothing like I experienced. The study's conclusions conflicted with the status-quo climate change crowd. So, the study began stepping on toes, lightly at first, but remember, I have enormous ogre feet so me presenting didn't help. I kept moving forward even as my confidence dwindled to initial reactions. I even verified the math over and over again under compulsory conditions. It all came back online when months after presenting the study I got a linked in message saying the study was brilliant along with two timber CEO's requesting meetings. Wow, that helped. But eventually both became a much longer story about logic. I was only providing excerpts at the time and usually built around AWC's.

The study then accelerated like rising global temperatures, fast and with all the pent-up energy of a teenager at prom. But I still backed away from defending it because I couldn't explain the logic, yet. Instead, I gathered opinions, although some thought I was promoting something. Far from it yet, the study didn't have the logic, just "it" facts that checked multiple ways and still strongly associated with AWC's. I'd hoped, I mean, I honestly believed someone else could save me from further climate study by already having the logic all neatly defined, but no one did. As far as I could determine, no one was even guessing at the logic. That really surprised me. I mean it all looked pretty-simple why wasn't there any academic papers covering it? There was one that came close. I got a lot out of it, and it appears in the acknowledgements. Still, I didn't have a clue then as to why it was missing definition from academia, but now I do.

Looking back, I knew the study was right and just needed to explain why it is. The why turns out to be important and was the undefined academic part that I could not find anywhere. I began to recognize the study was going to define something unusual. I described that earlier as everyone had a fleeting inclination towards "it," perhaps, but "it" certainly was not within their wheelhouse and in no way allowed by the current perceptions of climate change. It was clearly undefined, something entirely new to ponder. It was then I realized it was time for a separation of church and state. The study became its own and I shelved AWC's for the time being. It was a financial decision, pursue the climate study or risk the cash remaining to make a greedy company happy with AWC's. Not a hard decision for me at the time.

Solo it would be. Lone wolf, Olympic sprinter, a lost golf ball out in the tall, thorny weeds. Professionally, I was starting to get my ass handed to me for allegedly making some climate scientists look really dumb. The study could almost Forest Gump their reactions but that was never intended.

They weren't dumb. They all did then what we all do. We, including me, worked with what we had at the time, the time before this study, and that is working within "emission parameters." So no, the study is not saying or implying anybody is dumb. Not at all! Today, most have had enough time to absorb the study's new knowledge and accept it adamantly. Initially, I was left wondering if they would because, at that time, the study could only provide a partial, but still fairly complete picture but entirely without any economic motivation. It was really my fault; I'd asked too much too soon because the study didn't have all the logic to back the study up, not yet. Plus, no funds to pay for their privilege wasn't exactly a good plan to approach them with.

I could see something beautiful in the math and graphs, something unlike anything anyone had ever comprehended as a way to define and cure climate change. But there it was, staring at me from the screen and dancing through the notes filled with jargon while math's orchestra played a catchy and now well-modeled tune. The disco ball was lit and sparkling, but I was once again, not in the mood, and in need of a break. And now to make it worse, I was haunted. A real nightmare was taking form.

A nagging hurtful thought surfaced repeatedly. I had theorized something related many, many years before. A stupid naive theory that revolved around allowing forestry plots to grow longer to increase efficiency of harvest and lumber quality. That theory applied the basic and naive understanding that tree's get bigger with age and so does harvesting efficiency. That old theory lacked this study's much bigger picture of a climate cure, but thinking about it disclosed a simple factor that became universally applicable to now. Maturities positive effect on CO_2 sequestration. In hindsight, had I pursued that long-ago approach, I could have isolated this climate study's outcome earlier. The possibility of increasing maturity back then really stings me. As you read further, I think you'll understand why not following up becomes exceedingly unfortunate today due to time remaining on the climate game clock. Anyway, back to the story at hand.

Well, "it" was not so simple to define within the context of any other published findings. So, in the years before writing this, my understanding was like a third grader looking at a Van Gogh painting. The colors sure are pretty, but the underlying techniques are definitely not understood.

At that point, conducting the climate change study was becoming more annoying than IKEA assembly instructions, and there needed to be more parts, logic was missing. It was becoming a trudge because of the many learning curves involved. I'm not a climatologist, biologist, chemist, historian, economist, writer, or forestry specialist. A coworker once told me, that like him, I'm a "plain ole engineer." And Tom was right on that one. But that was what this study needed, fundamentals.

Real desperation began with a global search for logic, some way to define "it."

The study of climate progressed into accumulations of highly documented and referenced science done by scientists. Then I'd throw much of it into the recycling bin, but not all. One thing I learned in my career is you must rely on the works of others. It's the only way to accomplish big projects in a timely fashion. But I also learned first to check the data and credibility of all those "others." And do so thoroughly. Like most, I'd been hung out to dry after using incomplete studies and experiments in the past. Basically, science with non-reproducible results (low "P" value) that had been politically, institutionally, or economically motivated to deploy smoke and mirrors in order to sell buzzwords, catch phrases, and "transparency." So, I've been an embarrassed victim when young and eager. Now that I'm trained, that kind of science always brings out of me what my family calls my "Scottish thrawn." My dad was a Scot, a ginger with a fiery opinion and "bagpipes at the funeral and at the wake" Scott.

These days, we must wade through the swamp that reeks of half-truths and outright lies just to get a rare credible source to surface on the internet. Thanks to social engineering and AI, the internet is making morons into experts by the thousands, so long as they can put a few words together for click bait they can make money, so there is no shortage of them appearing over and over. But heck, they don't even have to make click bait anymore. Now AI and Bot's do it for them. To make the internet worse, I was finding myself outside of my education and experience and in huge learning curves. Solo once again. Phoning friends, public/collegiate academic journals, and new acquaintances helped a great deal in wading the swamp. But it really took a lot of time just to chase what the moronic influencers told me was my tail, and really wasn't.

But you only know good times when compared to bad times. Let's move on.

XPRIZE CARBON REMOVAL COMPETITION, BUT FIRST CHEERLEADER, NOT ME.

Now I can provide rosy and peachy comments to give you a warm fuzzy feeling about anything. I know how to do that. I learned how by seeing some real flakes hiding their incompetence with performances in hundreds of government engineering meetings. It's easy to stay cheerleader like and positive but exhausting if not your nature. I don't do it. Not unless it really is rosy and peachy. Then I love it too. I suppose I am me because I don't believe your stupid and I find false cheerleading highly disrespectful to one's intelligence. That kind of communication makes me really uncomfortable knowing its not fact based; it's a projection of falsified opinion to gain the audience's confidence. A confidence scheme. Anyway, I ask for you to bear with me while I introduce some facts that motivated me to perform this study. It's an attempt to further introduce myself so you can gain understanding of what makes the study really tick. My editor wanted to cut all this out and stay rosy. I didn't, so here it is, Scottish thrawn. The CMS struggle.

It's been my experience that if the slimy collective of no-profits were doing what they advertise I would not have had to research climate or written this book. They would have cured climate change decades ago with more global forests under protection. Not to mention, much of the donated tree replanting done over the last 50 years would not have been harvested. That's right, when it reached 30-40 years old your donated sapling was likely or is likely to be harvested. Essentially, many of the problems the study points out about climate might not even exist if only just a few of those entities really did as they advertise. It's a shame. Such a waste of all our good intentions and money should be made punishable. Those particular species of rat are in business to perpetuate climate change so they can continue to gnaw on it, not cure it.

Admittedly, there's good in many no-profits, but many do create lousy science to manipulate our friends and relatives with the PhD behind someone's name. You, as I do, may contribute to no-profits. And I am not opposed to getting involved with a proven good one. But here is something to consider about all of them. They did not know of this study when they made those nefarious decisions, did they? So maybe, understanding of this study now is what makes their previous decision nefarious. I'd like to think so.

Now, another no-profit caveat. And possibly, greatness exhibited by a no-profit.

I've entered into the Musk Foundation's, The XPRIZE Carbon Removal Competition. The judging starts a few months from now. I'm prepared in every way, sort of. I should have said despite how little money is available to CMS, we are trying. But even that isn't going to stop us. Why? Because the study's proofs are so firmly locked into Earth's future reality they can't be ignored, not anymore. But its new knowledge so we have to inform everyone first, so onto the XPRIZE global stage the study goes. We have to because it's a serious and costly challenge to spread knowledge. Its costly because internet companies own all information now and charge to inform anyone of anything. Even going "viral" now is tightly controlled by Google and social network providers. Yes sir, they own it and control it all these days. We gave it to them with all those "free" email and social networking accounts we all use. So again, it's back to money enough to even be able to inform folks of the study. And for the record, we ran a test with Google ads. Those butt heads charged us $47 for one click to our web site info page. That click was generated by a BOT and not a human. Yeah, total rip-off because they charge what they want to charge when they want to charge it with Bot's. They'd even activated our test ad without our permission and charged us $500 for the privilege and you guessed it $47 per click in a key word we deleted twice. Needless to say, quite using Google people unless you want them to do the thinking for you. Information should be free and not influenced by Google's bought and paid for view only. And who's paying Google's data extortion rates, well no-profits get huge discounts. Not cool.

So, methinks I might be understating the study's need for money part, greatly. CMS really doesn't have enough to do much of anything, not really even enough to ask for help. It's a struggle. No sense complaining about money yet, we just started, and I have faith in people. So, now to Elon...

Oh boy, some of what that guy says publicly is bothersome, but he doesn't always deplete my intellect and obviously he's trying to make some kind of a difference. Plus, I don't care about his individual opinions. They're his, not the study's. The importance of this study can't afford to care either. To explain, His actions with his XPRIZE Foundation and climate mitigation are truly admirable and that's the only link CMS shares. So, anyone can say what they want about Mr. Elon Musk, but that guy puts his money where his mouth is. Know of any other billionaires who are open to the ideas of others to make change (or charge)? Or uses innovation like the different XPrize's have? Nope, because there aren't any. Most billionaires use their foundations to patronize their egos, collect valuable artwork, real-estate, and promote their political agendas. It's a tax shelter thing, a blow horn for their opinions, and generally speaking not really about philanthropy. It's usually more about influence. Well, CMS needs that influence and advocacy that comes from winning the XPRIZE Carbon Removal Competition. And it can. In relative fact, it should.

Now, notice how I write "XPRIZE Carbon Removal Competition." Being in a competition, there are rules the study must follow. Writing it like that is one of them. Hmm? Branded like a cow in the 1800s. I wonder when the more modern ear tag will come out. A green one, please.

Humor aside, I hope the XPRIZE Carbon Removal Competition has some. At the very least, the Musk Foundation is addressing climate change and some other issues as well. The problem is, the study does not offer a sexy recent technology, so the chances are possibly slim. Plus, what company, like Tesla, wants to hear how they are wrong when it comes to curing climate change? Or even making a dent in it! Yes, so there is that. Perception and preconceived notions became reality after hearing the same story told too many times. Putting it down even after erudition becomes difficult. But again, they didn't know this study when they made it all up with what they had to work with, at the time. They now have an off ramp built from fact.

To me now, the people leading emissions-based carbon efforts, again stressing post this study, all seem more like profiteers gambling with human existence. I understand how those other thoughts generated. I also see the close association with Tesla's profit structure in European cap and trade developed. That relationship is the why they would risk the hundreds of millions in cap-and-trade revenue they receive and maybe have to pay into it to offset their CO_2 leakages? So yeah, there's that. But what a wonderful concept, their product becoming a true carbon net negative that actually cures climate change. Yes, that's an offer.

Dreams aside, I'm betting they won't voluntarily give up the millions a month in cap and trade to study findings, which means it might be made easier for the study not winning. Which leaves the potential of them saying, "Yes, we said we wanted to cure climate change, but we only want to use technology driven methods (which can't work because of physics, assuming they'll require energy input other than photosynthesis)."

I won't stop asking the public to help fund the study's mitigation efforts anytime soon. But I could be wrong in my personal assessment of their intentions. I mean it's more of a bad feeling then based on any facts or statements they've made. So, let's hope I'm wrong because it would not be only this study's win, it would be the world's 1st win Vs. climate change that ends 50 years of losses.

WHERE THE HECK WAS I?

Okay, back to the subject and much more specifically! I appreciate your patience in tolerating how the study came to be, how it shaped, and then progressed to here. It's time to move a bit faster.

The study pointed out that sequestration was still working, sort of, and had been working much better before humans engineered it otherwise. Fact is, it had been working splendidly for both Earth and animals for an extraordinarily long time. Like for hundreds of millions of years! But, now and all of the sudden, global sequestration is not working anywhere near as well as it should be. All that occurred slowly over eons. Human's tampered with and eventually broke terrestrial sequestration with our *demand driven forestry*. More on that term in a bit.

Learning that was a bit of a shock. It was a difficult lesson for the study to reveal because any in-depth study on tree maturity and its effect on climate just doesn't exist. Nor do studies on how much atmospheric ppm (parts per million) levels are affected by sequestration or how much CO_2 was coming out of the atmosphere during Earth's growth cycles, as outflows. Those all combined into a huge gap in climate knowledge that the study filled. Like the XPRIZE Carbon Removal Competition, it started to appear as humans are only looking for a sexy modern technology to replace what "it" already does and really can't be duplicated by humans.

We were looking in the wrong direction because the buss was coming from the other way. Time to get out of the roadway arrived. Contemporary climate pursuits mis an unavoidable and combined reality that keeps them from success.

Here's the deal, and the last time I'll mention this because you already know. Volumes of empirically verified material discussed "it," documented "it," and measured "it," and I even found volumes and volumes of publicly expressed empathy for parts of "it," all highly relatable to sequestration. Still, no person, group, or agency had added one tree and a trillion trees together to get "it"! Nor could anyone define the logic involved of why "it" mattered so much.

That was where "it" got exciting and complicated, where this study became taboo to the old guard of the dated science kingdom: the logic of sequestration and the beginning of thousands of science journal retractions and modifications. I can describe that better, I think, "we, and I mean all of us were wrong, and we knew it, but we didn't know how to admit that because we couldn't explain "it."" But now the study can.

TO START, TRY EXPLAINING "IT" TO ANYONE

Here we go, way back in the study's history, the first thing to consider when doing anything, the basics. AWC work documented the products using 12 amu for carbon weight, accurately modeling carbon stored within AWC products. However, weighing the carbon in the product (just under half its total dry weight) missed the big picture and really missed the enormous sequestration picture. All because of one question.

WHERE DID THOSE CARBON ATOMS FOUND WITHIN THE DRY WEIGHT OF TREES AND EVEN FOSSIL FUELS COME FROM? AS MENTIONED, PLANT LIFE. BUT THERE IS MORE TO TELL, GROWTH.

In all plant life, 97-98.9% of carbon-12 comes from Earth's atmosphere and nowhere else. Not from the ground the tree planted in and not from lousy science saying otherwise. Plants and trees get less than 3% of their total mass from the soil they're grown in. Their mass all comes from sequestering CO_2 from the atmosphere using photosynthesis. The plant's water, $H2O$, is extracted from the soil, which also carries a few other soluble nutrients up from the soil, but those nutrients help photosynthesis increase mass. So, and as I mentioned, not much of the mass comes from the soil, if any.

Observation of that empirical fact is extremely evident as tree roots lift sidewalks because they add mass under the sidewalk and don't deduct anything from the soil. Therefore, a tree's mass is a byproduct of photosynthesis, not magic, superstition, politics, nor internet influence. A growing tree or plant takes CO_2 directly from the atmosphere. In short, the plant absorbs, uses, and then grows from absorbing CO_2 from the atmosphere it lives in, not directly from the soil it's grown in.

That describes the entirety of CO_2 sequestration's role used within the study. Absorbing and breaking the CO_2 molecule down, then storing the carbon element while releasing the oxygen is another way of saying the same thing. The tree does this when the carbon is within a CO_2 molecule, not because a carbon atom has somehow miraculously appeared within its biomass or extracted from soil.

🌱 *We all recognize CO₂ as carbon dioxide, one of the vile molecules within our emissions and the stuff accumulating in our atmosphere that is most definitely causing climate change. Which is also causing the biggest existential crisis ever. But most don't recognize the amu weight difference in the CO_2 molecule and Carbon. Which is important to remember when it comes to sequestrations role.*

Here's a shortcut. Read on for details or skip to the bullet below to get the gist.

CO_2 has one atom of carbon-12 plus two atoms of oxygen-16. Simplified, $16 + 16 + 12 = 44$ amu (or, for precision's sake, 44.0095 g/mol). Also, one metric "tonne" is 2,204.6 lbs. A tonne of CO_2 contains 22,730 moles of CO_2 (1,000,000 g / 44.0095 g/mole). So, that's how many moles are in a metric "tonne," and the weight of one CO_2 molecule within that nasty tonne of CO_2 stuff. And so, what, right?

Well, "it" came into focus when I realized the difference in amu between the carbon atom and the CO_2 molecule. CO_2 is 3.67 times more in amu, or "weight," than carbon. As Yoda would say: "Chemistry 101, it is." Okay, maybe not. Weight is not mass, but it works for this example. Weight implies gravity, and mass is measured without gravity. Here's the critical part to understand.

🌱 *I came to understand that by reason, logic, and considering all the nonsense in this world, the following has, is, and will be an absolute truth, now and forever. "If a tree puts on 100 lbs. of carbon atoms within its annual growth period, it has sequestered or transformed 367 lbs. of CO_2 molecules to do so." (lbs. or kilos, it makes no difference to the 3.67:1 empirical ratio).*

That ratio is always going to be empirical in all possible ways of measuring the photosynthetic conversion in woody biomass. It will always come out the same no matter how accomplished. I combined that with my theory from when I was younger; that tree growth theory I didn't work but I should have. Well, the old theory now haunted me because of one fact it had disclosed, a bigger tree is more efficient and in every way.

To explain, I'll try and explain the 3.67:1's empirical nature to the study's context.

The 3.67:1 ratio expands to encompass one tree in a forest full of trees or one forest in a world full of forests, as in 10.8 billion acres of forest globally. Why? Well, a tree adding 100 lbs. of carbon-12 atoms within its annual growth cycle is nothing special at all. Some species, particularly in North and South America can put on hundreds or even a thousand pounds of carbon-12 per year. Some in Asia even more.

Here in Oregon, we have some species that are better at it then anywhere. And here's where it gets interesting. A tree or forest's maturity is what's most important to how much a tree or global forest can sequester CO_2. Maturity. There is no substitute for maturity in sequestration increasing its volume.

Maturity in any tree species is typically proportional to its size (it's mass). The older the tree, the bigger it gets and the more it has to sequester Carbon. Again, that's also empirical and common knowledge. A significance within this study arises because a tree in an arid location could take three times as long to mature its sequestration abilities compared to one in a rain forest. So, a tree at age 30 in the dry or arctic region could be the same or less in size as a 10-year-old tree in a rain forest. But all trees, at least the majority of tree species do, AND do so regardless of their geographical location, put on mass with age.

The percentage of mass added may decrease over time but the additional mass that percent accounts for is much larger yearly. Hence, sequestration increases greatly each and every year the tree is alive. That's because the tree or global forest is growing each and every year so it must support its previous years of growth while putting on new growth. Interesting as well, its mass and it's CO_2 sequestration ability is almost exponential in growth. So, maturity is critical to achieve both benefit's to forestry, as in mass and sequestration ability.

Again, I found a serious lack of studies on climate change that accounted for all the stuff needed, so nothing new to report there but the usual lack of data. Which I think I should try to explain with some speculation made pretty obvious by today's documentations. I think this might explain why science missed what this study turned up.

Most climate studies aren't concerned with sequestering (removing) CO_2. They're trying to reduce or even recycle it. There's nothing on tree maturity's effect on climate change. Generically, I found many, and I mean many, studies that weren't entirely wrong but were incomplete because they failed to examine or include "maturity," the 3.67 carbon conversion ratio, the principle of sequestration in general, and, all-too common, they don't equate sequestration and emissions to clearly define the balance differences. It seemed the world of climate science was missing many, and I really mean many, significant factors. Seeing the lack of data told me how right this study is. But hey, I want to be sure to provide credit to the many studies I used to figure out this one.

Don't get me wrong, when I talk about plant and forestry sequestration, there's plenty of data about both, and even well-established empirical measurement standards, with formulas that are beautiful, even gorgeous, and well used within the study. But they all trend toward micro studies, little pictures. I needed macro research, global-sized, or at least a complete study where sequestration, the 3.67 ratio, and tree maturity (among other details) all sung in harmony, like they naturally do. So, I began building the macro study using the data available which was plentiful just not complete, anymore. You're reading excerpts from some of that data now. But wow, it took a while to build this study from the incomplete ones.

One year and many months later, I'd found enough really weird and really good stuff to verify that the study had, yet again, done a thing for sure. And the motivation to keep moving the study forward. I could tell as I had a few conversations with grimaced friends and colleagues and heard their toes crunch under my big feet, again. Sure, they all came around after their toes and ego's healed and even became impressed with the work, unfortunately for me, they came around much later. BUT it was coming along! And once again, it wasn't time to celebrate. No logic yet...Darn it all! I still could not figure out why it all was.

No "Eureka!" yet. The excuse. Too early in the study. So again, I couldn't thoroughly convince colleagues. They could sense the gaps and attacked. They don't buy black magic without knowing how it's possible. The study needed more substance or the why, and how's. In short, the colleagues asked why and how? And I could only answer when and where. However, by then I did have more tools to build with, lots of them were turning up regularly the deeper the study went.

> 🍃 *And this is where the study takes off like a rocket away from contemporary knowledge of climate change. It's the toe-crunching stuff that's perhaps a little difficult to believe... at first.*

Past experience gained when presenting the study correctly informs the presenter that you will not believe a darn thing I'm writing very soon. "Experts" didn't. The smartest people I know were all dumbfounded for months, even embroiling me in arguments for months more. Heck, my wife, super smart offspring, and relatives all looked at me like I'd gone nuts. A couple of interventions were even employed along with some Voodoo in an attempt to bring me back from performing what looked to others like a career suicide. I resisted it all and then began noticing the firewood being quickly stacked under some lifelong bridges. But the study spoke to some things nobody around me knew; it was getting close to the finish. And to be fair to my people, they all came around when the hidden rabbit finally appeared from the hat. But it still took time even after. Please be patient from here on, pretty please with sugar on top.

ODE TO THE LOGIC OF IT ALL.

The study's flash of insight defining the logic came directly from Mother Nature filing a complaint. It occurred after compiling a crude Microsoft Excel model that actually worked the first time tried and exactly as theorized. Seeing it work repeatedly sparked me into taking a long drive into the woods so I could look at a modern logging site.

That site view produced an impression with the tears of me being so ignorant they etched into the climate concerns I champion today and until I die. It all hit like a freight train running over my primate mind. I did not like what Old Mother Nature had just disclosed. Humans were in for much bigger climate problems than currently believed. And unfortunately, the study had spoken for itself in easy to reproduce proofs. It was no longer a question to solve it was now an answer that must be dealt with, and it came with an uncompromising "or else."

Positive that I had all the logic needed forever burned into my head, but I still had problems defining the study:

sequestration's effect on climate change still needed words that didn't exist. At the time, the study's findings had become observable and measurable, but I could not explain in words and people reading my mind wasn't a realistic option. The problem was, and again, data about how it all combined to make climate change wasn't available, so the ways to describe "it" were also missing and would surely keep everyone but me from understanding the study. And that new responsibility to define appeared to be my own and my own damn fault for releasing the knowledge into our world. Basically, no good deed ever goes unpunished, ever.

SO, WHAT IS THE LOGIC BEHIND IT? OKAY, HOW ABOUT WHERE IT CAME FROM FIRST.

It all came down to an unexpected source: historical data on forestry that matched exceptionally well to historic atmospheric ppm levels (mostly from ice core samples). It was like they'd been traced from one to the other. So, years of my on-and-off dedication had just then produced all the precedence and proof needed from somewhere unexpected, history. The logic unfolded and I winced, terrified by what the study was yelling as caution. We are sprinting towards our demise and the study told me how. It was all a wicked plot, a trap.

🍃 *Climate change now seemed like a tripwire set by Mother Nature. She is now at war with humans and seems intent on removing us from existence. As a trap, she hid it by baiting it with what we believed to be a renewable resource, trees. We humans gladly took that bait and have now greedily consumed almost all of its maturity. As that trap is now closing quickly, we may have learned of it too late. Because trees were never the renewable resource we believed them to be. Unfortunately, we just didn't know and treat the forest that protected us from Mother Nature's climate trap badly and had for eons.*

The logic I found sparked that drive to the woods. And with my own eyes and for the first time I could see exactly what climate change actually looked and smelled like. There it was, human demise and within human form. Not a nuke, not emissions, and not some terrorist. It was all of us who owns or has used wood products. A clear cut, unrestricted logging, forestry demand. It all smelled of gaseous heat and looked like a war zone with its stumps and shredded limbs. Knowing the study and solo again, climate change was no longer existential, it had become a depressing reality in possession of a physical oddness. It is no longer invisible; it has a presence. I now viewed immature trees within an otherwise perceived beautiful forest with such a distain and disappointment in being human I cried. Forest views could no longer provide euphoria, that was replaced with an agonizing regret for being involved in their now; study recognized demise. The study's new knowledge has filled me with regrets I'll carry to my end. Especially from that concept of years ago, the one I didn't follow through on, man it really haunts me now. I should have. I could have. But I didn't, and that would have provided more time to fix it. As a result of me failing, the study outcome can only say "maybe" to the fix it fast enough question.

NOT THE WIN I'D HOPED FOR. SOMETIMES YOU GET FEATHERS WITH YOUR CHICKEN.

And yes, I had finally done it. Except now I wished like hell I didn't know anything about climate change. And certainly, I never wanted to know what it looked and smelled like. And sure, the math worked, the evidence was in, and finally, the logic defined. I ended up coining 11 or so terms to describe the logic, *constrained deforestation* and *binary restricted resource* being the most significant of them, and voila! Eureka! At last! It quickly became a piratic victory to human awareness, confounded in memory, bleak in any future. Still, the study was the first to describe exactly what caused climate change, so victory had arrived, no matter the cost. The study also tells us how to fix it, which is good, but "to be continued."

Man, I really did a thing for sure and what have I done? I would have been better off not knowing and traveling with the wife, playing with the grand kids, and dying off as expected. That is no longer a future knowing what awaits my offspring, it can't be.

Since I was supposed to be happy with finally defining the logic, but really unhappy with what it all meant, I forced myself into realizing something else: "it" deserved a name that would define it as a scientific discipline regardless of the dismal possibility of the future it also revealed.

I began to refer to my work as Complete Mitigation Science, or CMS for short. And with that my attitude improved because the study does say we have a chance to fix it. But we can't dilly dally about. There are possibilities worth taking.

The possibilities became clear as I continued. Opportunity sparkled like a silver fishing lure to a hungry trout in a kid's fishing derby. The title then grew to the most significant description possible. "Complete Mitigation Science of Atmospheric CO_2 from Residence Conditions and Emissions, Complete Mitigation Science, CMS."

It's a real mouthful, but the title is accurate. What CMS defines is the simple beauty of sequestration benefits to Earth's biome. The climate-changing puzzle is solved. Complete Mitigation Science provides the once-missing pieces. Eureka again! (this time with more enthusiasm!)

Okay, maybe that wasn't very modest, but how else can I say it? Possibilities! Suppose you combine the CO_2 benefits from all climate mitigation attempts out there. In that case, CMS blows their combined totals away like they're all a climate-melted birthday candle in a hurricane. Sequestration works and has been working for millions of years. It just needs to be repaired with maturity, all thanks to us. And that's the trouble because there's no other way out of climate change. CMS can't be ignored.

Now let's cover more important things, like staying alive here on Earth. That under good circumstances is difficult enough. But it can be impossible when mother nature has been angered enough to set a "renewable" IED (improvised explosive devise) for humans or if you need human's to make a good group decision.

BUT WILL HUMANS USE IT?

For a few minutes let's forget about the study's proofs that are a page or two away. Or just skip the next two chapters if you'd rather learn CMS proofs and findings first. You can always come back and learn what the biggest problem is for CMS to cure climate change after learning what CMS actually is and does.

Will humans use it? Probably not. And that is the Greek tragedy that explains the problem of curing climate change soon enough. It's frustrating to see the airplane heading into a mountain while you're yelling, "Pull up!" to the drunk and passed-out pilot! To keep it from happening we all must understand why the pilot is drunk in the first place and allowed to fly the plane. It is easier to let it happen then become aware and stop it. Just ask the mothers of many criminals, they still love them regardless of who they hurt. Climate change is in that same rut of denial. We love our toilet paper, paper towels, fast food containers, furniture, homes, and a lot of stuff that comes from trees. More on this later, because with CMS it doesn't have to all go away. Meanwhile...

CMS is not a sexy, flatter, giant TV with a clearer picture, a shiny electric car that's newer and prettier than your neighbor's, or a more brilliant phone that takes better photos or videos of your lunch, vacations, or kids. Nor is CMS neatly wrapped in foil, a beautifully pictured tax-deductible no-profit. CMS is a foul-tasting antibiotic with crappy side effects that you could only get from a doctor's visit. But it eliminates the misery of that sinus infection the kids and school gifted so nicely to watch you suffer.

In short, CMS fails to appeal to the more sociopathic nature of our society. Emissions-based so-called "solutions" do appeal to that side of humans which is why they're popular. For the record, the appeal of emissions reductions has proven successful at seducing humans, not curing climate change. 50-plus years of failure prove that. CMS has no instant gratification like solar panels saving you money or trendy electric cars that look shiny and zoom, nor does it hand out anything for free. CMS requires the work needed but not the work wanted by our natural proneness to do as little as possible to get by. CMS requires effort.

But CMS has many things going for it. I mean wow! It can really fix climate change. It can save us from certain destruction, even extinction. That should take priority in our lives, right? But they don't. Likely, CMS won't either. That is, until it's too late.

The police showed up after the crime. Politicians change policy only after an incident that more than likely came from the previous policy. Humans go to the grocery store after running out of food. My point is, more often than not, humans react instead of being initiative-taking. Which is why humans must change our perceptions into the actions needed done, or else.

That is what this book is attempting. Change the perception of climate change of every human possible: make Earth a better place for the future than how we were given it. Now let's hit the first hurdle to curing climate change.

INTERNAL PSYCHOBABBLE: PERCEPTION

Your first impression of CMS is likely that of me being a quack or some kind of con artist.

I see the perception problem every day, just like you do, pathetic pathways designed to entrap us, to fool us, to force us into believing absolute horsepucky. Those numerous everyday traps are what make promising ideas harder to get past initial perception, even when they don't come from the stereotypical "we want to rip you off" message that is devoid of logic and fact.

Perception issues with CMS started with me while I was reviewing its initial results before I defined the logic. I was given the first opportunity to cringe and then struggle to accept it. I suppose it was a privilege. No, no it was not! Old Mother Nature's flash is not something you can unsee. To put it bluntly, I found the results hard to believe even though the math checked out and the logic eventually became rock solid. At that time years ago, I did what any human might do: I lost confidence in CMS. I fought against proven facts. I rechecked the math and begged a few peers to check it too. I reviewed their reviews and checked again to include all the logic and correlations along the way. It was all sound. So why had I entered doubt's grasp as some do now?

I'm not a shrink, but I recognized what was happening within my brain: a subconscious debate. All these new, better, contextually defined data streams conflicted with what I'd been taught and had comfortably accepted. Emissions ruled. So, it was easy to ignore it all and take the kid to Disney World for a graduation present. But I thought about CMS the entire trip. Even after the TSA helped someone steal my iPhone and wallet in North Carolina's airport security check-crooks. Anyway, all that not-so-right data was deeply ingrained, and it was being constantly reaffirmed in media everywhere I went, so my perception of any kind of sequestration-based climate change solution dulled, and then gave me fits. All because "it just couldn't be so simple!" Even though it is once the study defined it all.

That's the basis of the perception problem that faces CMS and the issue of solving climate change today, that darned internal conflict we all go through to change how we think, especially after we've been proven wrong. Beating perception will not be, and is not, easy to do, because when it comes to advancement, human egos consistently construct failure out of thin air surrounding greatness. CMS is set for that sort of perception failure, but let's not let that happen by giving ourselves some time to absorb new knowledge. So far, it takes a month or two for people to accept CMS, you're probably no different. So, from me personally, see ya in a couple of months after you put this book down and think about what I'm writing from here on. Just don't wait too long to come back, we need you.

TIME TO TEST PERCEPTION

WHAT DID NOT CAUSE THE CLIMATE TO CHANGE?

As it turns out, emissions are a problem, but not the entirety of it, and the Industrial Revolution wasn't when climate change started. Emissions are an input, but they lack precedence in actually causing climate change. Yes, there is no denying that CO_2 emissions are accumulating in the Earth's atmosphere and helping climate change kill us off. They accumulate because they have no other place to go, but they used to. I know, that's far from today's thinking, but it's the truth. And so exceedingly difficult to readily accept. Which is the problem. So, humor me a little with another attempt at explanation.

Humans cannot avoid releasing CO_2. Therefore, CO_2 has to have a place to go, sequestration is where it needs to go because ultimately, that's where it came from. Not the best explanation, sort of vague, I know, I'll do a lot better a little later.

You see, the more people learn about CMS when exploring its proofs, the more its perfect logic takes hold, and not because of how I say it, it's because of what it says, because it's not opinion or hearsay, its fact that you can replicate with many methods, not just one. That's all I have: facts. Unfortunately, these facts are antibiotics needed for survival, so follow the directions carefully.

ON TO A SMALL SHOW OF FACTS. LETS GET IT STARTED!

The first one: everything I just said about emissions and the beginning of climate change is both accurate and can be initially hard to understand, and now easily proved. The next one: some 80% of the global population knows climate change exists, which is good. The bad part of that is it isn't anything like what they "believe" it is. The remaining 20% doesn't care, doesn't know, or is somehow motivated otherwise. Those statistics make you critical to fixing climate change, but here's the next problem I made pretty obvious:

CLIMATE CHANGE IS NOT WHAT YOU BELIEVE. NO WHERE NEAR IT.

You probably think human emissions are causing it, so reducing emissions will fix it. And your struggling with that earlier perception test. Time to make it worse. Climate change was actually geo-engineered, unknowingly, over thousands of years by humans using forestry resources. We've all been somewhat but unintentionally manipulated into thinking about emissions and the Industrial Revolution as the cause of climate change. It is a common knowledge thing that is not accurate but widely accepted. Here's where I grab your hand and take you to the playground to meet some brainy friends with laptops and graphing calculators.

> 🍃 *Climate change began with human domestication efforts tens of thousands of years ago because it (climate change) is not solely dependent on CO_2 emissions. It's more respondent to CO_2 sequestration abilities (as fast-cycle CO_2 terrestrial sinks).*

Therefore, the beginnings of climate change is summarized as both *sequestration and emissions dependent*. The two must, at the very least, be in balance for even modest homeostasis. However, because CO_2 sequestration relies on atmospheric residence durations and random photosynthesis global sequestration ability is more consequential in creating climate-changing conditions than human or natural CO_2 emissions can ever be. Natural emission levels are highly related to plant and animal quantities on Earth. Human emissions are just bad and unnatural to Earth from the burning of fossil fuel. The key word here is "conditions" That separates this study's scope on sequestration from the quasi norm of emission-based reasoning. Why? Well...

🌿 *Climate change is caused by human forestry demand, not by natural or human emission levels. It is an environmental impact brought about by geoengineering forests over eons. The impact is measurable as an "impeded fast-cycle CO_2 sink," as the impeded sequestration effect.*

The geoengineering of forest's is from the improper and inefficient uses of forestry resources during past and present human domestication. Specifically, decreased forest recovery durations negative impact on forest maturity. This well-pronounced environmental impact is measurable within a current tree or forest age when you consider what that age could be, had it not been harvested. Since maturity dictates current and past CO_2 sequestration ability that age difference also measures the level(s) of the study's *impeded fast-cycle CO_2 sink* as in how much it is and how long it has been impeded.

When you compare a tree's sequestration capacity now to when its sequestration capacity is measured with human absence the impeded level can be millions of percentiles. IE: A tree absorbed 140 lbs. CO_2 one hundred and fifty years ago and was harvested that same year, if the tree had been left alone and never harvested, today it could have been absorbing 3,000 lbs. of CO_2 annually and increasing yearly. Or roughly for this example, about 2 million percent (%) impeded. Yikes!

It is so unfortunately true it hurts. And that scenario provides a proof about Complete Mitigation Science.

> 🌿 *Therefore, if you remove humans from our historical forestry use but keep todays human and natural CO_2 emissions levels, there will be no such thing as CO_2-driven climate change.*

That particular proof's scenario and its outcome assesses the scope of the study. A colleague of mine loosely suggested setting it up as a fairly simple experimental model to check the study's credibility. I say loosely because his intentions were to disprove CMS. The opposite occurred because the model worked as CMS predicted and is 100% repeatable. In scientific terms, its results make the CMS study's precedence undeniable. That is one of the ways peer reviews works to check results, answering objections with facts.

This particular result really honed the CMS point and really got my people to stop the interventions and put away the Voodoo dolls. It got them all thinking sequestration. And me? Well, when asked, they all know me again.

Unfortunately, proof that cannot be denied seems to prompt the old guard of academia to become offended. It's really not what anyone wants to hear because the old guard has been saying something entirely different for decades, emissions. So, it's hard to hear a good message when I'm throwing mud in your eye? Yes, you could say it like that because it's true. We were all wrong about climate change, to include me. Start printing the retractions old guard! Let's hope they do so without being asked by promoting CMS knowledge to all those young eager minds out there. I can at least hope.

I HAVE SOME EXPLAINING TO DO

> *Before I can drive these next points home let's peek at how trees sequester CO_2. We'll then get into how demand driven forestry affects sequestration to make our climate change.*

I'll quickly explain what global *fast-cycle CO_2 sinks* are. These are terrestrial CO_2 sinks, so they are land based. Mostly, they consist of trees within forestry and really include all plant life that absorb CO_2 from the air around them, photosynthesis. Yeah, we all know that. More to the point. Earth's fast-cycle CO_2 sinks rely almost entirely on global forestry in order to function. That function is intended to remove vast quantities of CO_2 from the atmosphere during Earth's plant growth cycle(s).

Sequestration, in its simplest definition, is the amount of CO_2 any given tree/forest absorbs through photosynthesis, allowing the conversion of a CO_2 molecule into a single Carbon atom into its mass (size). Which we all probably know something about. Going further and interesting relation is constructed.

Trees use carbon dioxide-CO_2 and expel oxygen-$O2$. Which is the opposite of animals, which breathe in $O2$ and exhale CO_2. Trees absorb the C from CO_2 and release the $O2$. Humans and all animals absorb the $O2$ and put back the C to make CO_2 again. It's a symbiotic relationship. As in one is needed for the other. It took Ma Nature hundreds of millions of years to build that relationship and about 7,000 years for humans to screw it up. Our ignorant audacity is breath taking, isn't it? We do love working without a plan and Engineers like me are hated because we make people use plans. But they are not always good plans.

Our side of that relationship make unnecessary demands on our partner. That is because everything that has anything to do with human domestication releases CO_2. <u>EVERYTHING</u>. A step further and a clause in the prenuptial agreement humans have with Earth. Earth also releases CO_2, like humans. Earth's CO_2 emissions are called "natural emissions." And wow! Natural emissions are 10's of times more in volume than human emissions and good reason for the prenuptial disclosing them. Natural emission's come from animal respiration, volcanic activity, ocean vents, natural decay, and are really too numerous to list all sources. What it boils into is Earth's and Human emissions are in fact, the same within our symbiotic relationship with sequestration. Earth's sequestration arrangement doesn't care where the CO_2 emission comes from, only humans, incorrectly, do. And that is a violation of the prenup. Thats a deep thought, isn't it? Well, I can go even deeper. We can't have one (in present condition) without the other? Again, humans are more *sequestration dependent* and here is how the study revealed that insight.

- *During a typical 30-year-old tree's annual growth cycle (spring-fall), it will absorb, use, or sequester from the atmosphere some 163 lbs. of CO_2. Now, the crucial part: At 72-73 years old, it can do the same to 1100 lbs. of CO_2. Quite a difference, right?*

Say what? At this point, I'm hoping your intellect sparked. The 1 + 1 = 2 sort of stuff does that to me! Complete Mitigation Science is all about having older mature trees. If only it were that easy. That kind of kills the moment let me try again.

Okay, maturity increasing a tree's sequestration ability is both incredible and empirical in measurement. But it's not time to celebrate yet. We have a problem working against us to use that information and cure climate change for good. So here we go again, jump back to move forward.

CLIMATE CHANGE STARTED LONG BEFORE THE INDUSTRIAL REVOLUTION.

The study explained that climate change isn't so much an emissions problem as it is an environmental condition. Yes, again it's a "condition." To explain, let's visit a long time ago in human history.

Let's begin with human domestication, when, millenniums ago, we began settling into fixed places after developing disciplines like animal husbandry, buildings, and farming instead of remaining nomadic. As time progressed, we got better at using resources like forestry to enhance our lives and progress even further.

Before, during, and since, we modified our surroundings to thrive, and wow, do humans do that well! Eight billion of us is surprisingly definitive proof of our success. To accomplish so many of us simultaneously occupying the same planet, humans worked with what humans had, could beg for, borrow, or even steal with warfare. We figured out how to use resources. But it seems not so effective as we thought.

Forestry was no exception to human domestication's principles: humans used almost every bit of it many times over. Using it isn't the problem. Our plan was not so good.

The problem is how humans have historically cared for it. Stewardship is discovered needed in order to enhance sequestration in forest renewability discussions. Thus, improved stewardship is needed to eliminate the historic forming of climate changing conditions. And ensure forming those conditions in the future doesn't happen again. Now that we understand those conditions create climate change; they need to be discarded in the practice of forestry stewardship.

Later, I address a separate subject and provide a proof of when climate change actually began. I'm keeping it as a surprise to provide a much higher impact when it's needed then simply dropping it here. I also need to highlight the following to clarify that revealed later. Plus, I repeat myself too often already. So, for now, the study addresses the thousands of years of human forestry digressions finally catching up with human domestication. We'd gotten away with the abuse of our forested biome for thousands of years, but Mother Nature presented a bill to be settled around 1850 CE, or else eviction proceeding will ensue.

- 🍃 *Historical forestry demand and its correlation to atmospheric CO_2 ppm levels serve as CMS's focus and create a "datum" on climate change's impact starting.*

I use "small" here as a comparison to population levels of today. Really, I could have used "Tiny" or "insignificant" to describe nomadic population levels. I think you'll get the point...

Being nomadic meant small populations could only develop. These populations lived off of the ecosystems that could temporarily support their numbers. One reason, maybe even the most significant reason, why humans changed zip codes so frequently as nomads: they'd use up all the resources in an area. The game stopped migrating through their area another, but the game became less of a reason once farming and raising animals became prevalent. Any way you look at it, moving happened frequently with our predecessors, and even up to the early 1930s dustbowl farmers. It still does in some tiny corners of the world, which my urbanized ape sized brain finds peculiar but sometimes appealing when my family backs up a drain at home. I'd rather move than plunge it out.

By nature, humans are still nomadic. At least I am. If the workplace, grocery store, and gas station all close or a war erupts, humans move elsewhere without much thought. We may not like it, but we will relocate to available resources or safety. Humans don't stick around when the area's resources are used up. So, we don't wait to go bankrupt, starve, and then get drafted as cannon fodder unless there's good reason to fight; humans relocate.

The thing is, being nomadic is better for natural environments like forests. That can be pretty obvious over time as whatever was used up regrows or reproduces. But it is not so good for the people who had to move constantly to stay ahead of slow regrowth and animal reproduction timelines. What I mean is this. When our predecessors left an area, they did not return for prolonged periods, if ever. Staying away gave the used-up place time to recover naturally. But not having steady resources, like we enjoy today, also meant not having the large stable populations that resulted from making babies, lots, and lots of them, convenient.

Progressing in human domestication, humans now have limited nomadic conditions. Today, humans interact with nature intensely, not passively as nomads. Our natural resource intensity has proven a successful survival tactic: 8 billion people are again the proof. Nevertheless, that intensity results in problems to work out, having geoengineered climate-changing conditions with forestry use is the largest of them.

Arguably, at the top of the intensity list is how humans currently use, and more importantly, manage, forests' resources. Right now, humans manage forests for their biomass alone. We did that to meet the demand for forest products. The study calls that *demand driven forestry* practices. Unfortunately, that stewardship model is highlighted by the study as only fulfilling half of current human needs. Human domestication requires both forestry biomass use and now more than ever forestry provided CO_2 sequestration.

Historically, forestry stewardship missed or negated the dual role of forest renewability. The study defined that miss as *sequestration dependent* and trees as a *binary restricted resource*. The fact is, 85-95% (or more) of Earth's forests today are used to meet *demand driven forestry*, entirely ignoring sequestration. That percentage really took off around the CMS-established *climate change datum* of around 1850 CE. It seems to peak around 2000 CE, yesterday really. The study's datum ties it to what happened historically and continues today with what causes CO_2 driven climate change. A chapter on the *climate change datum* appears later where we get into more detail. For this introduction, it is all about the CO_2 of course, but in what way?

Humans have "successfully" modified our forests to suit demand. First, there is that word again: "demand." To insure we're both on the same page, "humans want or need it, so humans get it." Making forestry answer human demands came with geoengineering forests. Which is considered an achievement in human domestication. And it is... sort of. I said "sort of" because we know that part of the forest is easily renewed. In fact, we've proven that many times over by regrowing trees previously harvested. But our methods to regenerate income from forests were placed in much higher priority than regenerating sequestration and overall forest efficiency. That bad planning has pointed and accelerated humans towards an almost certainty of extinction. Now more than ever, a statement I heard about ruining the planet and killing us all for a few coins comes to mind. And I don't remember where I heard that, maybe a song?

TIME TO SHAKE THE TREE TO SEE WHAT FALLS OUT. WE GET TWO RENEWABLES BUT IN NO WAY EQUAL.

Here is a point the study makes that common sense seems to govern. That said, this could be why nobody saw Ma Nature's tripwire. And you're only going to find the CMS study saying it. Why that is, I haven't a clue.

CO_2-sequestration ability is not an easily renewed resource even though it's part of the proven renewable biomass of the tree. A tree or forests' CO_2 sequestration is limited due to the time required to mature and allow the combined resources to become net carbon neutral or negative. Again, the study coined the term for that as sequestration is a *binary restricted resource*. The production of woody biomass as the only renewable is where our forest stewardship is now proven to be failing all of us. As the study points out, trees aren't the renewable humans all thought they were or in the way we treat them either. Literally, the study points out they are not the renewable as they are currently being advertised and never have been.

FOLLOWING UP WITH CAUSE AND EFFECT

Here is the "what is" that is making CO_2 driven climate change conditions. You must see, the 30-year-old tree given in the maturity proof, it will likely never reach age 40. Not within today's *demand-driven forestry* practices, it for sure won't. It's harvested as "efficiently" as possible via clear-cutting somewhere between the ages of 25 and 40. Trees are clear-cut out of the forest regularly. It's rare for ANY tree in any forest to mature beyond 40 years. And given the flakiness of data available on the maturity subject, it might meet a much earlier demise as low as 30 years old. Sometimes they can make 45-60, like during economic downturns, recessions, depressions, etc. But generally, 30-40 years is all they'll mature to.

But surely not all forestry falls into the timber reproduction category, right? I wish I could say no, but that would be repeating a terrible half-truth told by no-profits, captured science sources, and some timber companies, who didn't know better. The United Nations provides some global forestry data that is helpful in untangling fact from fiction from many tainted data sources. But always keep in mind where even the U.N. gets data from, or from whom, meaning they didn't count the trees or do the surveys. They are more or less publishing timber producers numbers.

So, the verifiable truth is difficult to verify. However, the study has isolated statistical ranges using multiple sources. Unfortunately, we know that somewhere between 85-95% of forest acres globally consist of marketable trees harvested around age 40. And somewhere between 12-17% of global forests are protected and contain roughly 3-5% of untouched mature trees, the rest are immature or not forested.

To be confident in later checking the study's results, the study used the lower end of all the ranges. So, 85% of global forest land is used to model. That keeps any error possible very conservative, really any error becomes nonexistent because as you increase the range to say 95%, errors diminish to nothing. As you'll see later, the truth may be as high as 97%, but there's no conceivable way it's lower than 85%. Yikes, right? You better believe it!

And here's where you can do an easy study pretty quickly. Pull up Google satellite imagery and tour around the world. See the non-planted forest land, trees so small they look like grass, with truly little of anything appearing as old or mature growth anywhere. Just compare anywhere to the Amazon Rain

Forest or a U.S. National Park (with trees) and you'll find the contrast visually easy to make out.

How about that? You and Google maps just made a CMS study proof!

> ❦ *It's important to be aware of something about CMS's relation to available data sets. It's been a struggle to find and isolate sequestration data simply because sequestration isn't the primary pursuit of mainstream climate science, emissions are. Sequestration is a new idea, a scientific breakthrough, a new discovery, or however you want to say it. As I write this few know of sequestration's value to Earth, and to human domestication.*

CMS and the maturity requirement it points out is not the focus of forestry data providers, economics is. The foresters are currently focused on how to grow trees faster so they can harvest them sooner. Forest product engineering is looking for ways to process even younger trees, so investment returns faster. In addition, a well-funded, academic, in-the-way, not-for-profit, government-funded, climate-destroying effort is called "Mass Timber." The Mass Timber program is why this study was funded privately; Mass Timber sucked up all available money.

Curing climate change doesn't fit Mass Timber, bio char, or bio-burning electrical-plant grant announcements, and that really is all there is today for grants that are "close" to CMS. Nobody expected CMS to show up with an actual cause and the cure. So, for the time being, Mass Timber and similar projects get awarded to Fortune 500 companies paired to no-profits and to universities wasting valuable resources and time, right now. They got all the grants because they worked the system for years prior. Billions. You'd think CMS would have gotten something from its grant applications, but it couldn't, because it wasn't the right political flavor, yet. But we've found advocacy among grant agencies, so the study is becoming more popular every day.

Meanwhile, CMS still gets pigeonholed easily by bureaucrats because it's interdisciplinary. Which means, unbelievably, there is no government grant category for "climate cure." Like categories named biochar, Mass Timber, and the other politically connected buzz words. You'd think "climate science" would have that government sub-category but it doesn't. So, when the study does a grant application it doesn't fit the parameters, regardless of what it does. My ape brain struggles when logic can't be applied, is yours?

It seems climate science is looking for ways to reduce and measure emissions and is generally studying the insignificant, as dictated by grants and budgets written by people who are trying to do good. Engineering in general is looking for a technology that can't exist because if it could exist it breaks all known laws of physics. Most historians are paid to study wars and political soap operas, not forestry use. Politicians are looking for ways to live with climate change while getting insufficient or modified data to suit current agendas. And government research funding, well, get in line and bring political advocacy. The research caper is intended to augment the well-established bureaucracy found everywhere these days. Oh boy! The best and brightest are not in charge of anything!

And what have tax dollars funded besides 50 years of failures? Renewable and reduced-emissions energy programs. Programs that leak carbon before, during, and past their useful life spans. And zero sequestration efforts, well, in fairness CMS is the only sequestration program that makes any sense. What has been awarded to small companies' and small researchers amounts to hardly anything and comes with so many strings attached you need that miniscule grant just to manage the red tape costs. I know I'm griping too much and there is something positive to bring-up influenced by our failed attempts.

Cap and trade. That can be a very beneficial element to CMS and curing climate change, but it cannot do so with the standards currently governing its ongoing experimentation in Europe. It has a few kinks to work out, but CMS sequestration standards make it work as it was intended. Plus, CMS can openly participate. Enough already, lets get on with it!

It's time to introduce actual data. To do so I've added some of the graphs the study used to help in analyzing CMS.

Our first chart, *Figure 1*!

Owner class/ land class	Region			
	U.S.	North	South	West
	Million acres			
All owners	766	176	244	346
Timber land	521	167	210	144
Reserved forest	74	7	4	63
Other forest	172	2	31	139
National Forest	145	12	13	120
Timber land	98	10	12	75
Reserved forest	27	1	1	24
Other forest	20	0	0	20
Other public	176	35	20	122
Timber land	63	29	15	19
Reserved forest	47	5	3	39
Other forest	67	0	2	65
Private corporate	147	29	65	53
Timber land	111	29	61	21
Reserved forest	0	-	0	0
Other forest	36	0	4	32
Private non-corporate	298	100	147	51
Timber land	249	99	121	28
Reserved forest	0	0	0	0
Other forest	48	1	25	22

Figure 1 complements of: USDA, Forest Service 2014, *U.S. Forest Resource Facts and Historical Data, FS-1035.*

Figure 1 is valid data on who owns and controls forests within the United States. Full disclosure, this data is not as valid as it is referenceable; it'll do for this point. The truth in the forest industry is it's all about the money. That will probably never go away when it comes to forestry (unless governments use eminent domain and end climate change, that is. Sorry, sometimes my dreams come out in written words. No money can be made, and nobody reelected when applying eminent domain!). Never going to happen and that is good! Still?

Okay, let me point out in Figure 1: 766 million acres are forested within the United States. 521 million acres are on the "soon-to-be clear-cut again" list. That doesn't account for the "other forest" category, 172 million acres, but you can bet those trees are or have been marketable in the recent past. Therefore, 90% of the United States' Forest, 693 million acres, has been and will again be marketable, or, if you prefer the less politically correct term, has been and will be "logged" again.

So, if it can be logged, it's in the rotation to be logged. Few exceptions exist to that economic rule that governs global forestry under *demand driven forestry* practices. But we can fix it, with money of course, we just have to pay for a lotta trees. Easy right? Maybe. But now, let's jump across the pond.

Compared to Figure 1, the rest of the world is in the same boat and generally speaking worse off. Say, for instance, Russia. They have difficulty finding a tree big enough to make a single wooden matchstick. Some even say they have to glue one together from saw chips before they can sell it to China. Speaking of China, they should have trees, lots of them, but they don't. China also bought Russia's over the last three or four decades and then sold them to the world in products. Their construction boom and economic expansion consumed and is still consuming a lot of wood. I give these examples to stress just how deprived global forests have and are becoming because I could have said everywhere instead of Russia and China.

Now, let's talk about the data used in Figure 1. It's validity—the reserved forest of 74 million acres. I call malarkey on that. Much of that reserved forest was logged and will be logged again because most of those acres are managed by the Bureau of Land Management (BLM) or the U.S. Forest Service (USFS is part of USDA), and, you guessed it, have been and will be logged again. Laws that make logging on public land happen also force those agencies to market trees, even if they don't want to. So, I say the use of "reserved" to describe those acres is wishful at best. Because:

The 52 million acres protected by law as national parks and national monuments are not on the chart. Guess why. Some 70% of that protected land isn't forested, and what is forested was likely logged before becoming protected but logged over a century ago and not since. Therefore, there is a shiny spot made by National parks and monuments that's not part of Figure 1. Those protected acres aren't managed by BLM or USFS either; plus, many countries have National Parks (but not near as big as the USA! and Canada!). Those national parks and monuments with old growth trees really do help sequestration. I've got a demonstration of just how important those protected lands are coming up. But first, maturity.

AN ESSENTIAL COMPONENT OF THE STUDY, GLOBAL TREE, AND FOREST AGE.

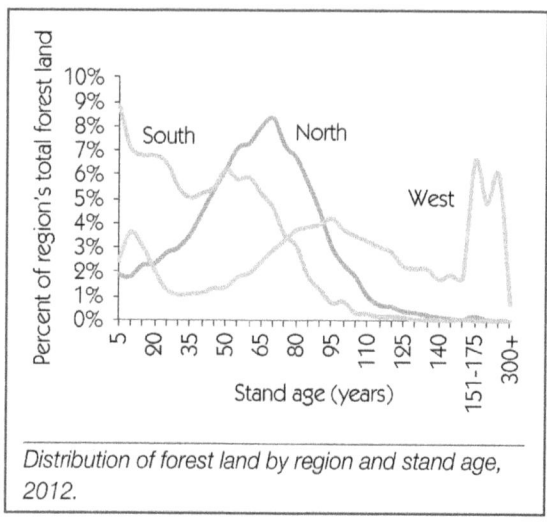

Distribution of forest land by region and stand age, 2012.

Figure 2 complements of: USDA Forest Service 2014, U.S. Forest Resource Facts and Historical Data, FS-1035. Each of the three lines represent a region in the United States as North, South, and West. The line(s) represent stand age within the percentage of forest in the indicated region.

Figure 2 clearly indicates Forests over 300 years of age only make up less than 1% of the United States' forests. The 1% of remaining old growth is our national parks and monuments, plus some tucked-away acreages that are probably too mountainous to log or lack access. Combining Figure 1 and 2 tells us that up to 90% of the United States forest has been or will be logged again between age 30-40, (average stand age in Figure 2 data is around 40). (Now, wait for it!)

But hey, you'll only find old growth in two places on Earth. One is in the United States, so hurray! We did a good thing! Thank you, President Teddy Roosevelt. The second is mostly in Brazil as what's left of the Amazon rain forest, which is a lot more than 1% old growth, but only because it's not cost-effective to log yet, plus some laws and international agreements trying to protect it more. But still, it's going in the right direction, so hurray! Now to the bad data caveat. (Stop waiting, here it comes)

Figure 2 graphically demonstrates an earlier point I made about data but be careful: the Figure 2 graph is entirely constructed not by me, but by those with the intent of deceiving the reader. It conveys a highly tainted but easy to believe message. And fortunately, it is really easy to discredit by reading into what is not there.

The smoke and mirrors used is explained like this: Figure 2 is only scaled up to 10% and started at 5 years old. So, if you look at where the south line begins, just below 9%, or any line, know this: most tree stand ages are left of the percent (%) axis and off the percent scale deceptively provided, meaning they're either less than 5 years old or treeless lands at the time of the survey, both of which are the actual majority age and percent of land. So, evidence of the deception is they didn't start the graph at age 0, so nothing younger than 5 years old is on it, and there are certainly no just replanted or treeless clear-cuts (bare ground) on it either. That would make the figures look worse and seriously decrease the percentage of forests at a given stand age. Meaning, where they say 8% of stands in the North Forest are age 65, yeah right! It's more like 1 or 2 % at best. Being scaled to ensure those negatives weren't represented is appalling. The message sent is they believe we're fools. That's my initial take but there is something else to consider.

That makes one of my points about available sequestration data. I mean, taxpayer-paid scientists concealing tree ages? It can't be, right? Seriously, what's their motive to use a deceptive statistical practice? I hope they just lacked the data, and that could be true. What else could it be? Perhaps capture? Okay, capture is when a government agency ultimately serves the industry it is formed to regulate. Yeah, as I mentioned before, the data they get is what they use. So, it's likely just a case of laziness or lack of funding. They didn't do the survey or bother to fact-check the timber company data provided, nor expand it to actually represent reality. Now get this. Figure 2 is released annually by the Forest Circus, I mean Forest Service, USFS and has been in this format for decades. Write them an email and tell them they are &*$%! for doing it. And here's why...

Finding a 100-plus-year-old tree globally (outside the Amazon or National Parks and Monuments) is like finding a matching sock in the dark. If there are any in any numbers, I certainly don't know where they'd be, except for those sticks that look like miniature trees growing out of the permafrost in the arctic north. Only because they're not marketable, too small, even though they're really old, their environment keeps them from getting big, which probably saves them. Anyway, overseas, they're lucky if they have 30-year-old replants because wood demand is so high. Europe and Asia both love their stuff made from wood! And then there's something else. The more dictator-like a government is, the less maturity there is in that country's trees. The corruption effect, or communism, on forestry resources is also sequestration's enemy. Use Google Maps to look at Vietnam, North Korea, Russia, China, Iran, anywhere there is a dictator, communism, or widescale corruption you'll see their forests are very immature if presentable at all. All though, Vietnam and China are making an attempt with reforestation efforts generally those efforts are economically motivated and not intended for sequestration enhancement. I'm not going to mention what

Malisia did to its forests or the Philippines, too depressing. But then again, they didn't know.

To bring tree age into today's context, here's a not-so-historical point about tree age. Democracy isn't any better with tree maturity. We all love the stuff that comes from trees! And worse has happened more recently, other than Mass Timber that is.

You probably know about the devastating fire in Paris's Notre-Dame cathedral. The fire destroyed most of it. The good news is that France didn't hesitate to rebuild the historic landmark and has made a lot of progress. The unwelcome news is that they should have rebuilt it with steel and not wood. Why? Because they scoured Europe to locate and then cut down 1400 Oak's that might have been the last European trees with any maturity, of any quantity, to make the timbers it required. Sad but true. They had to scoured the land to find the older trees used. They needed their huge size to make huge timbers. They didn't hesitate, and now those trees are gone for another 140 plus years.

Which makes a contemporary point on the study's tree maturity finding; had they known how important those trees were to climate change, perhaps cutting them down might not have happened? Losing that many old trees is for damn sure not worth the climate damage that CMS now demonstrates as really, really horrible and comes with consequences France likely feels right now. But again, they didn't know they were accelerating our extinction to rebuild a historical/religious attraction. With this book, they do now, and I can only hope they feel as terrible as I do about the years ago study, I should have completed.

Below are links to the Norte-Dame information:

1. https://www.ewc.company/downloads Our CMS site has both sources article's.
2. https://www.smithsonianmag.com/smart-news/dozens-century-old-oak-trees-felled-rebuild-notre-dame-cathedrals-iconic-spire-180977481/
3. https://www.npr.org/2023/06/11/1179648233/notre-dame-paris-fire-rebuild-roof

SO, HERE IS THE PROBLEM, RIGHT THERE.

- Societies' economic rules dominating forestry demand keep trees from maturing, not just a few of them, almost all of them. Economics creates *demand driven forestry.* The greed within that scheme readily exchanges forest and tree efficiency gained with maturity for the quick coins gathered from significantly less efficient forestry management. Basically, the forest industry takes the few steel penny's made right now in exchange for the many gold bars made tomorrow. History proves that is a stupidity only found within greedy or ignorant decision processes. And that's what's managing, for the most part, the entire global forest today, greed.

The U.N. estimates that over 2 trillion trees exist globally. Wait, that's good news, right? Yes, it's good we have trees, but again, their too immature to do any good. Plus, the increasing number and intensity of forest fires due to climate change is also related to tree immaturity. The world forests are producing evergreen matchsticks, canopies close to the ground and surrounded by brush. And now Russia... I'm not picking on Russia... Okay, maybe I am. Anyway, Russia has over a trillion trees themselves. Too bad they're mostly less than 20 years old. Even worse, substantial clear-cut lots where trees used to grow absolutely cover Russia. Have a look on Google Maps but bring a hanky to dry the tears.

Here's yet another alarm sounding. The majority of global trees that have managed to become adolescents, the ones that sequester 100s of pounds of CO_2 annually by themselves, are, on any given day, in a logging rotation of some kind. Those trees are 30-60 years old. Any tree over 30 is sorely missed by global sequestration standards because it just became a net CO_2 negative when it was harvested.

The clear-cut site those trees were growing on immediately begins leaking CO_2 into the atmosphere. It takes 20 to 30 years after a clear-cut for that area's replanted trees to sequester more than what's being released. That is what I call the *carbon hump*. It takes even longer to become net CO_2 neutral if natural regeneration (allowing adjacent trees to seed the clear-cut portion) is used to regrow the lot. Natural regeneration doesn't cost money, although it does require significantly more time, as in decades, to regenerate the biomass portion. The cost makes it the preferred global afforestation method.

IT IS WORSE THAN YOUR THINKING.

Knowing tree immaturity reality sucks. Let's continue the downward spiral with one of the study's findings on old growth. So, the good news is the remaining 3-5% of mature global forestry sequestration is the only thing saving humans from a complete climate collapse. That's a bitter pill to swallow but still good news, isn't it? I'm really only talking about the Amazonian forests. The old growth they contain provides the only real braking mechanisms against that collapse. Only 3-5% remaining globally should be downright terrifying but take refuge in knowing Earth still has "some" measurable fast cycle sink ability elsewhere; but not near enough. Plentiful sequestration is a rarity, away from the Amazon Rainforests and national parks it also comes from crops, grasslands, and older trees everywhere are contributing, but the way-too-numerous immature trees aren't doing anywhere near what's needed or what they could provide, and as it stands today, they never will. Again, CO_2 has no where else to go but cause climate change.

When the study revealed that, I began to wondered, "Is it too late?" Maybe not. That's the only answer to provide for now because of the "Will Humans Use It Question." Currently, the truth about the state of Earth's forestry sinks is limiting our odds. Right now, our survival is not a good bet even for the riskiest of gamblers. What I mean is I used to worry about nuclear war but now that I understand sequestration's role nukes have become dated. Anyway, beating climate change isn't a given, now that we understand what it is. Now, it is going to require luck.

Your help is needed to change Earth's luck, that's our shortened game plan. The longer game plan. Suppose humans can expand the remaining mature trees by allowing the younger ready-to-harvest acres to mature. In that case, the brakes on climate collapse will improve significantly every year. That is precisely what CMS mitigation curbs do.

Eventually, CMS, if used and we catch a break or two, will stop climate change, and make the world a lot nicer place. I'll show you the math on that in a bit.

To start CMS, we must be careful which acres and trees CMS uses to gain the quickest results because our initial and extremely limited funding won't give a lot of room for error. And here I am dreaming again, funding. What funding? That stuff is needed, anticipated, and making me old fast.

And of course, the forest acres adopted by CMS must be over the *carbon hump*. That limits the supply available, so we have to go global right away and find as many as possible. The older the trees, the better. Surprisingly, the effect on climate will happen rapidly and because of CMS that too is measurable. How about that? The ability to actually measure CMS's progress, well it's true, and we can right here and right now, empirically.

A PEEK AT SOME RESULTS ANYONE?

Protect the old growth that we have first. Old growth is around 1.3-1.8 billion acres globally. It's also the first part of figuring out how large a forest is needed to supplement what old growth does already. So, no more Notre Dame rebuilding for a couple hundred years please. Old growth sequestration provides the base needed for CMS mitigation curbs to expand. We need to mature more trees into old growth to improve sequestrations volumetric efficiency. Surprisingly, not as long is needed if we can acquire trees over the *carbon hump*.

Without getting into numbers coming later, a large-enough 30-year-old forest that's allowed to age another 10-15 years can significantly impact climate change by adding to current old growth sequestration volume. Get enough 30 plus year old forests into CMS so they can become 50-80 years old and CMS stewardship can absorb all human and natural CO_2 emissions made. So how large is large?

I rudely answer that with a question, "how much can we get?" Well, we can get a lot with a lot of money. Or nothing with no money. I'm sarcastic because that is the answer whether we like it or not. However, there is a trick that applies in our favor, *proportionality*.

Immature trees require more acres and more trees while old-growth trees need fewer acres and fewer trees. That also applies to the time needed, younger trees need more time and older trees less time. With that trick our goal is to achieve a balancing act between atmospheric CO_2 ppm while more immature trees are added into the mix while others become older growth and decrease the acres and time required.

To broaden my attempted explanation, eventually, a few postage-stamp-sized tree plots on the toy Earth globe in your childhood classroom added to old growth acres of today is all it'll take. Spreading them out across continents would also help. But it'll take time to get there with trees growing a lot slower than humans. But then again, there is a lot more trees then humans! But when we do accomplish the proper proportions of tree maturity, those postage stamps on the toy globe will negate current, future, and all past emissions from CO_2 driven climate change. So, it's definitely worth working towards and won't take as long as you might think. As I mentioned, enough trees maturing, and measurable results are obtainable almost immediately with very serious climate change reductions within 10-15 years.

The study show's sequestration is powerful stuff and can add to our luck. Unfortunately, humans have limited Earth's sequestration ability by millions of percent, so curing climate change will require many acres initially but not as many years as you'd think. After all, we already have 1.3-1.8 billion old growth acres (mostly) doing what the study pointed out, so, adding acres should be easy, yeah right! Anyway, the more acres we can get initially is all the better. And more old growth obtained and protected means fewer acres. That's the *proportionality* trick the study provided to end climate change. And we know it works because it did for hundreds of millions of years before humans messed it up.

Finally, I would love to see a law that does not allow a tree over 60 years of age to be cut down, anywhere in the world. Regardless of its ownership. Doing so would keep things like Notre Dame from happening and would also force thinning around older trees that could actually improve that trees sequestration. Unfortunately, there aren't a lot of that age anywhere so it wouldn't be enough soon enough; but it would be a step in the right direction. Oh, and the punishment for cutting one down, well, torture then a slow painful death involving the tool used to cut it down. You may think I'm joking, but I'm not. As you'll see in a bit a tree that age is rare and priceless, as in no amount of money can replace what it does to stop climate change from killing us all.

DEEP DIVE WITH A WARM BRAIN

FIRST, AN OFF-RAMP BUILT FROM TIMBER

I await a phone call from a timber management company. I've been talking with them for weeks. It could be the first full-scale CMS stewardship operation. Here's an update. "They want upfront M-O-N-E-Y, darn it. I really thought they'd invest. I'm sure I told them CMS is just starting and has a tiny budget, okay, so they either don't give a dinosaur's care about that big burning thing in the sky coming fast, or they need the money. Either way, I'll keep trying, elsewhere if necessary; maybe they'll come around eventually."

"It appears I need to raise coins like rabbits. And okay! All of you were right! I'll write the book for yet another attempt at fundraising!"

CMS's mitigation in reality and its despair are distinctly different things currently occupying the same space. So... "Timber Landowner, here's your off-ramp to more money with efficiency. Some upfront sacrifice required though. You must be willing to save the world." End of CMS advertisement.

And on to the next timber company I went, which eventually will work out, so, we're now implementing, our start, and I did write this book. Now, I just need to figure out how to turn all these darn rabbits into coins!

DATUM—WHERE IT ALL BEGINS TO REALLY START TO SUCK

It's time to the dive into the CMS data pool. Starting with the *Climate Change Datum*, I call it the year 1850 "ish." Honestly an actual and extremely specific date is in flux. However, the study's most recent NOAA ppm data assault is coupled with forestry usage data and then connected to some secret sauce in historical events that used a lot of trees and then accounted for modification of land use for spice that got us to 1850ish. But for the record, I can't get us to an exact year without a supercomputer and a couple of gifted people to model it. But I sure would like to. It has to do with only being able to work with 255 data points at one time whereas I have 1,000's of points to analyze. Yep, a supercomputer and a couple of nerds are needed.

With that disclosed, after much deliberation and a dartboard filled with fact-driven intuition, I settled on 1850 as being close to, if not the year that climate change consequences started to hit Earth hard. Note: this is not the year climate change started. I've got a presentation on its start a little later.

What can be said accurately about 1850 and why I chose it? The certainty of 1850 as the datum for CMS comes from the correlation between historic forestry demand and atmospheric CO_2 ppm both taking off. Those happened because *constrained deforestation* had begun spreading around the globe and by 1850 it was really prevalent globally. *Constrained deforestation* practices continued to intensify after 1850 and showed in the atmospheric ppm data as it spread further each year.

CONSTRAINED DEFORESTATION

As opposed to the typical definition of "forest degradation" where forestry is expected to regrow into a forestry normal, CMS's "*constrained deforestation*" is forestry that is physically kept and not allowed to achieve a normal level of forestry and CO_2 sequestration by *demand driven forestry*.

This effect is typically due to human interference from inefficiencies, commercial harvest rotation timelines, or non-human influence with unintended forest impediments brought on by humans, weather, fires, climate change, and biological events. The primary act is human's cultivating immature trees that are never allowed to mature it's *binary restricted resource*, sequestration.

- 🍃 America's, I'm saying North, Central, and South America. I put this explanation in because I've gotten many questions that made me look it up. The term is commonly used to refer to all three geographical regions but not as often as it used to be.

Almost everything that could be easily logged was logged as *convenient forestry* at least five times and as many as fifty times before 1850, and it has all been logged many times since. After 1850, the Amazon River basin and the west coast of the Americas were getting clear-cut. In short, the last old-growth stands had become marketable, and humans were incredibly happy to cut them down for a few coins. Concisely, the America's were now being fully indoctrinated in European and Asian *constrained deforestation* practices and sequestration volumes plummeted as the result.

THE AMERICAS' WERE LOSING THE LAST OLD-GROWTH FORESTS TO DEMAND DRIVEN FORESTRY.

The America's inland and coastal forests weren't considered easy to get to or process because of their huge mature trees in very remote and difficult to access locations. Which makes me sad thinking about it, hindsight being what it is. To give you an idea of just how huge those trees were, think about this; one limb off of old growth as equal to two or more trees harvested today. These huge trees were not *convenient forestry*. Still, railroads, ship building, and cities were all growing like mad from 1800 to 1930, and all that growth required the last remaining untouched old-growth forests to be delivered to the sawmill. Trees were delivered from everywhere, including from convenient parts of the Amazon, some of which had now entered into harvest rotations while other portions were being converted into farm and ranch land. That turns out to have converted 38% of all global forestry that has yet to be returned.

All the world's previous forestry use by 1850 had tipped the scale in climate change's favor, but it is nothing like today's much higher intensity. It took a century after 1850 before the majority of global forests ran out of old growth and were replaced with the immature tree rotations of today. That was accomplished between 1950-1990 and then really noticed in climate change ppm measurement when the Amazon's old growth began declining from 1950 to today.

The 1850-1920's railroad construction made *demand driven forestry*'s expansion possible. Railroads needed millions of rail cross ties, lumber to build themselves, boards to build towns and cities, and firewood for their steam-driven equipment. Inevitably, rail lines began to extend their reach into the vast old-growth forests of the Americas which brought those forests to marketable status. The world's great railroad expansion used every stick of wood it could get. I'm not kidding. Almost all of it. Everything steam-powered before 1850 used wood as the primary fuel until coal replaced it at a higher cost. It only replaced it because steam power ran out of *convenient forestry* from which to acquire the better and more useful wood fuel cheaply.

 That decision was forced onto humans by Ma Nature. And thank the powers that be that we didn't ignorantly strip the Earth of old growth trees back then, or else, climate change, per CMS, would have wiped us out already. Our existence on Earth really is that delicate and we still might have learned that fact too late, time will tell. Back then, we really dodged a bullet for sure, and with complete ignorance! I hate to say it, thank goodness for, what? coal? No, no, no! I just can't say that. There is way too much wrong with coal to give it kudos. Secretly, "thanks coal, you saved humanity, at least for a while!"

By 1915 CE, even the remotest global forest was made marketable by railroad/steam expansion. Around then, gasoline-powered trucks also started logging. Trucks didn't have the up or downhill restrictions nor the expense of building railroads, so they quickly became preferred. Trucks could inexpensively go even deeper into the mountains and extract old growth without the expense or restrictions of building railroads or being required to float logs in rivers.

The higher impact created by trucks was vast as new roads were constructed to take advantage of the technology. Trucks are a major part of the Industrial Revolution that started a century before, 1750-1900 CE. Which did accelerate forestry's sequestration demise, so the Industrial Revolution was a contributor to climate change but the geoengineering of it all started 7000 years earlier. Towards the end of the Industrial Revolution's timeline is when the global forest stewardship failure was paid to really zoom.

LUMBER'S NOMINAL MEASUREMENT SYSTEM, ARTIFICIAL DEMAND.

The early 1900s is where the timber barons joined together with the U.S. government and promoted the first timber "conservation" program, called "Nominal Measurement." This program was far from conservation but highly advertised as such. It is devastating to global forestry. This program and others are not unlike what's happening today with the government-funded Mass Timber program, making it possible for smaller and smaller trees to be logged and turned into wood products. It's all done to create artificial demand so return on tree investment is faster.

The modern practice of lumber measurement uses a nominal measurement scale. Nominal measurement can be called greed, bad forestry management, or what it really is: answering the public's demand for forestry products with *constrained deforestation* to generate profit faster.

As the study concluded, nominal measurements historically create artificial demand as one actual measured 2" x 4" lumber component (load-calculation determinants) is equal to 1.25 to 1.5 multiples of the modern nominally measured 2 x 4's as actually measured as 1.5" x 3.5". That simply means to do the same job or work (measured as force or newtons) you require not just one nominally measured 2 x 4, you need "two" of them to substitute for one 2" x 4" board. But here's the scheme. When you buy two to do the same job, you actually buy more board footage because 1.5 x 2 is 3 and 3.5 x 2 is 7. So instead of a 2"x 4" board you bought a 3" x 7" board to do the same thing. Sneaky isn't it.

Nominal measurement perpetuates impedance of forestry fast cycle CO_2 sinks by restricting recovery durations to the resources biomass return only. It also occurs within ALL global forests used for contemporary wood productions. It is a global practice that provides no alternative due to its wide public acceptance and over a century of engineering accustomed to the lesser performing, lower quality boards. Plus, log sizes needed to reestablish an actual measurement in component don't exist anymore. Nor does the tooling required to process log sizes to reestablish actual measurement or quality. So, we are stuck with it for what might be forever. Unless of course things change.

Nominal measurement's artificial demand had increased forestry demand by at least 30%. It also made profit per board foot of lumber almost double. The additional profit made timber producers at the time into the "Timber Barons" of today. Those profits paid the way for those Timber Barons to land grab much of America's forested land over the next century. They still own those lands today and with those lands comes life or death like power over the global biome. They are now in charge of Earth's Thermostat.

Another proof to the study appeared within this comparison; "which had the larger impact on climate change, fossil fuel emissions or nominal measurement's artificial demand?" Even the impeded sinks observed during nominal measurement implementation show the value of sequestering CO_2 from the atmosphere. Sequestration diminished much faster as nominal measurement implemented and appear regardless of increasing CO_2 emissions of the same timeline.

SO... Referring to Figure 3 later in the book. Although the trend that magnified in 1850 as CO_2 ppm's were increasing before nominal measurement, the fast cycle sinks are still establishing notable peaks and valleys but are increasingly in more noticeable retrograde as nominal measurement implemented. Before and during that time, the growth cycles basically demonstrate some remaining sequestration potential; although, they are obviously struggling in an impeded existence before. But after the added strain of nominal measurement occurred it really amplified their failing trend to not viewable on the figure. Therefore, and again, fossil fuel emissions lose in comparison to the sequestration depletion created by nominal measurement's artificial demand. Sequestration is again determined more relevant to climate change then emissions.

WHEN ENOUGH IS ENOUGH

The nominal measurement thing really brings out the Scottish Thrawn. Should Timber baron's choose to ignore CMS, you'll know exactly who to thank for climate change's impacts. And for the record, I am certain none of them knew of or wanted that responsibility. But if they react negatively to that now well-established responsibility, they should be treated no different than mass murders, cigarette companies, or serial killers. They will quickly become enemies of the state, enemies of the world, and an enemy of humanity. It is therefore necessary to try and assist them in adjusting their profitability with CMS stewardship first or lay the blame entirely on their shoulders while extracting reparation.

- To be sure, humanity is stating a zero tolerance for anyone acting within a scheme that now so obviously includes the demise of an entire planets biome, creates difficulty in domestication, and accelerates human extinction.

I can't help reminiscing about how the oil and coal people are villainized by climate changes emission-based sciences. The reasons they were never tied to the stake and burned is possibly because we're painfully aware we can't live like we do without them. I also think their misinformation campaigns, science for hire, and contributions to academia to come-up with plausible deniability worked too well. Those stratagems worked because emission-based science is and has always been questionable because it wasn't really provable. One, because 50 years of making no impact speaks volumes. Two, look at CMS's proof about removing humans from forestry and keeping human emissions. Emission science could never prove anything like CMS has. Emissions science was and still is a theory and not fact. CMS does not have that problem in either defining what caused it or how to cure it. So timber company's take heed.

You will not be able to deflect responsibility like the fossil fuel industry has. Ironically, and I hate to say this, but it's no longer the fossil fuel industry's entirety. Wow, that was really tough to admit. What I mean is, now they can only be charged with the lesser crime of human emissions in the global trial of the crimes against humanity. Timber companies get the lion's share because sequestration has proven far more critical. Saying that makes me cringe. First, timber didn't know what they were doing, it wasn't intentional. Two, I'd grown to despise fossil fuel emissions and all it stood for. COP28 made my head spin like I was possessed (oil company sponsored and in charge).

For the record and again, fossil fuel use is bad and needs to go away with clean energy solutions. But not all things fossil fuel are bad, 8 billion happy smiling people on earth is all the proof we need of that. The bad parts need to be dealt with but that's not the scope of this book. Forestry and sequestration are.

THE CONFESSIONS IN PPM DATA

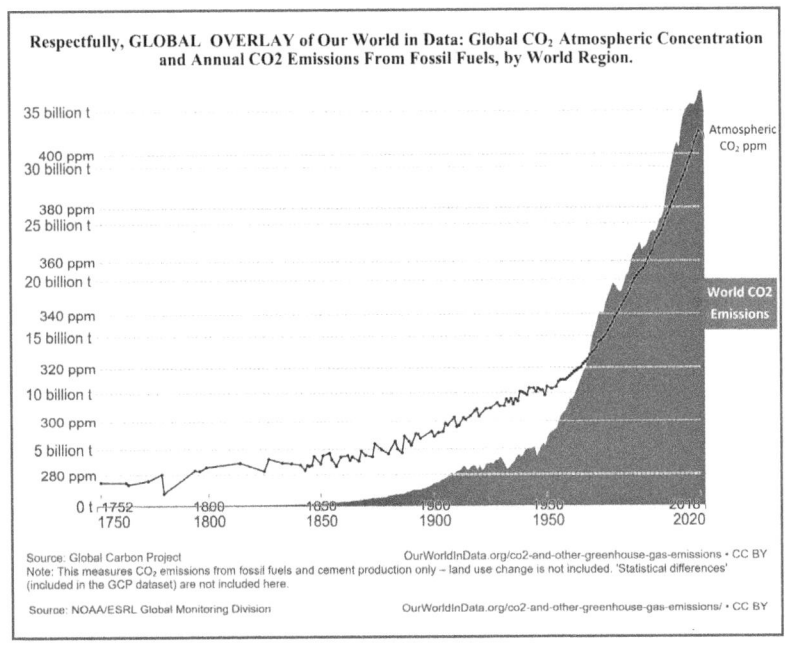

Figure 3: A picture of 1850 in CO_2 ppm, 1950 ppm, and Human CO_2 Emissions. 1752-2020.

This book's small format makes some things difficult to see in Figure 3, but you should be able to see that climate change existed long before human CO_2 emissions took off. Figure 3 also shows the peak demise of old growth up around 1950 and nominal measurements impact just prior. The chart leaves only one conclusion: sequestration was being affected because emissions had not taken off. The ppm and emission lines may look similar but they're unrelated. Don't believe me? No problem. I initially saw the same similarity between the two as well but have a look at Supplemental Figure S1 in the back of this book. It provides a clearer and longer-term picture. It also fulfills the previous foreshadowing of, "when climate change began."

Have you taken a look at Figure S1? Good. Millenniums of rising ppm and almost zero human emissions, hmm? What do you think now? Figure 3 is just a zoomed in picture from Figure S1 so both Figure 3 and Figure S1's ppm levels correlate to forestry demand. Much of the ppm graphing acts like it was traced from historic forestry use. But here's this book's problem: it is a "cart in front of the horse" thing that needs addressing.

I can't release the correlation graph as of this writing. I'm still adding to it as more globally related historical forestry demands are isolated. Language barriers slow me down because much of the historical data are in foreign languages. It's also more wallpaper-size than book-size, so putting it into a book currently seems impossible. I'll get it onto the website when it's peer reviewed. It'll be free to download. I'm also including some papers on the subject covering the topic like Palestine cedar forests demise, Roman and Greek deforestations, Roman deforestation of Germany and England, deforestation of Spain and France to build the Spanish Armata and English/American global ship building efforts, global railroad expansion deforestation, Global forests to farms. You'll get the idea and see the timelines.

> 🌿 *The study's cited documents and other graphs too large for the book are on the website at: https://ewc.company/downloads*

Be sure and subscribe while your there!

We don't need the big correlation graph to make the point. That's because Figure 3 and S1 speak volumes. To start, around the studies 1850 datum there's a unique sequestration signal in the ppm data, the first among worsts. You can see it made a plateau 1850-1851 basically two similar highs were achieved. It's the first year that the global CO_2 fast-cycle is seen jumping further out of balance with emissions and begins signaling today's climate problem with its accelerating trend. That negative to the climate indicator worsens from 1850 on. And you can also see the fast-cycle sinks shrinking in numbers and ability from there on. See the ppm upticks and down ticks decrease from then on, until they literally disappear around 1950 (at this scale). None of that disappearing fast-cycle or growth cycle pattern relates directly to emissions, but emissions affect the 1850 date and one of the reasons a supercomputer would come in handy.

Okay, have a look at Figure 3's ppm line before 1800. That upward movement before 1800? That's volcanic activity, and I mean a lot of volcanic activity in the years surrounding 1800. You can see a huge CO_2 inflow around 1760-1790 than a small outflow. Interesting how it then begins stabilizing (moving sideways). I currently theorize that sideways movement indicates a balanced CO_2 inflow with outflow 1800-1820 or so, which seems to indicate the 280-ppm level was maintainable by sequestration at the time. That balance was lost around 1825 as *constrained deforestation's* global expansion continued. For now, know that volcanic activity created a lot of pre 1850 chatter that needs filtered out to perfect the study's *climate changing datum*. And from 1750 to 1825 even volcanic activity couldn't keep sequestration from doing its job. There is a reason for that coming up in the form of a proof.

THE PLANETARY GROWTH CYCLE

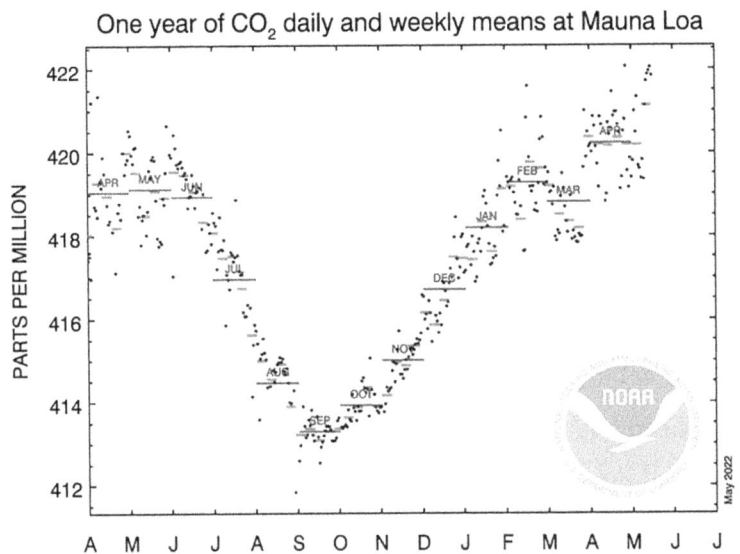

Figure 4, Mauna Loa CO2 ppm Measurements.

This Figure 4 should make it easier to see the sequestration signal disappear from the 1850s to the 1950s in Figure 3 and S1. Okay, it's time to demonstrate a typical fast-cycle sequestration pattern: that is our planet's growth cycle. To do that I introduce Figure 4, *Mauna Loa* CO_2 *ppm Measurements*. Basically, a snapshot of what Earth's growth cycle looked like in 2020-2021. Figure 4 was made in 2022 and shows a typical fast-cycle sink/growth cycle of Earth by measuring CO_2 ppm levels weekly. As the name implies the measurements were done at Mauna Loa Observatory in Hawaii. The black dots are the CO_2 ppm readings taken during the indicated month. Note, the ppm trend ends a lot higher than where it started. That's the runaway I'll be talking about soon enough, unfortunately.

An average global growth cycle within the fast cycle sink pattern is an uptick in CO_2 during winter, then a down tick during the fall season. It's the hill and then the valley you see in Figure 4. That pattern existed as Earth's plant's growth cycle long before humans did. It did, however, begin to be influenced negatively around 6000 BCE as seen in Figure S1. By the study's 1850 datum, it started to crash faster and has almost disappeared today. Unfortunately, the measurement of that trend has gone negative twice in the last seven years, first in 2016 and again in 2019. The evidence for this is available as "CO_2 ppm Delta's" in the free CMS manuscript off the web site. It's also too large for this book's layout. Ultimately, what the study found in ppm measured growth data is not a good thing for us, not good at all.

To explain, after winter, spring arrives, and the natural sink cycle should reduce atmospheric ppm's during growth periods—meaning spring to fall, which happens every year in a big ppm swing, or at least that's what should happen. The previous Figure 3 provides a big picture view of the growth cycles' upticks and down tick trends. But note Figure 4's 2020-21 up and downtick is not viewable in Figure 3. Thats because the graph's inflows and outflows disappear in the buildup of CO_2 ppm in atmospheric residence and one other more important reason. Their disappearing. You can see they are diminishing in amplitude 1850-1950. In later years we have to zoom in to Figure 3 to even see the cycle because of the year to next year attenuation in their recorded signal. Meaning, their going away because they are becoming weaker over time. And that is a problem. A huge climate changing problem.

Before moving on. Do you see the down tick from February to March on the right-hand side of Figure 4? Good, here's what that's all about. Earth has two hemispheres, so we also have two growth cycles, a slight uptick occurs in February and a slight down tick in March, first because the cycles unhealthy so its small, and second; only 32% of Earth's entire landmass is in the Southern Hemisphere. So, what you're looking at is largely what's left of Amazon's old growth's sequestration doing its best to create an outflow, but it can't. I would have mentioned the other Southern Hemisphere countries, but there isn't enough old growth to really measure. But I must!

Specifically, the Southern Hemisphere down tick is created by roughly 90% of South America, the southern part of Africa, and the Australia/New Zealand sink cycle. Their growth cycles are during opposite months: summer in the Northern Hemisphere is winter in the Southern, and winter in the Northern is summer in the Southern.

Before 1850, the cycles trend was indicated at the end of every fast-CO_2-sink cycle (by fall) was just above where the CO_2 ppm began at the end of the previous winter. Meaning a lot more sequestration was happening before and even around 1850 then now. To give you a general summary of the idea, the CO_2 ppm outflow trend line is around $y = 9.56$ CO_2 ppm sequestered from the atmosphere from 1974-2021, but that's declining. And it's 30-35% less when compared to before 1850. Before 1850 there was almost enough sequestration ability to deal with natural and human emissions, combined. But the previous to 1850 damage is also notable, even millenniums before. In a nutshell, up to around the 1850 datum, it all worked "well enough" that climate change would have taken thousands of years to show up as it did around the study's 1850 datum. Which segues into another critical point of climate change and study proof.

- 🍃 CO_2 sinks don't mind needing more CO_2 than they can get. It's more natural for plant life to lack CO_2 than to have too much CO_2. Too much CO_2 is known as excessive CO_2 fertilization, which creates plant growth problems.
- 🍃 By this measure, it becomes obvious that Earth should always have more CO_2 sequestration ability than CO_2 emissions as its norm. Currently, Earth is both unnatural and extremely opposite to that position.
- 🍃 Demonstrated in Figure 3 and S1. The years around 1752 and 1790-1820 with sideways moving ppm levels around 280 ppm indicating somewhat of a balance. In comparison, the increased upward trend beginning around 1850 indicates an unbalanced sequestration to emissions ratio.

CO_2 upticks and downticks only trending up, up, and up again didn't make it hard to pick a starting point of a climate collapse beginning around 1850. Yes, climate collapse. Thats what really began around 1850. Around 1850, climate change impacts were noticed by Mother Nature because of that upward ppm trend developing firmly from there on out. She'd had enough and began to react politely then, but today, not so much. That occurred as the old growth CO_2 sinks were disappearing even quicker, which threw off the hundreds of millions of years it took her to balance sequestration with emissions.

CO_2 SINK CAPACITY IS ALSO ESSENTIAL TO CLIMATE CONDITIONS FORMING OR BEING ELIMINATED.

This is the section where I explain what the study really dived into. It started with a theory. That theory looked at the contemporary understanding of the global growth cycle and used atmospheric ppm measurements to confirm its validity.

The theory, "The contemporary understanding of global sequestration today is that it performs, functionally, the same task annually. Therefore, it should be measurable in year over year trend as a baseline or an extreme detectable variation within atmospheric CO_2 outflow volume, this measurement should be almost a static volume in growth cycle episodes." To explain, atmospheric CO_2 outflow volume should remain consistent year over year. If outflow volumes remain consistent then too much inflow (emissions) are the atmospheric accumulation problem. If outflow volumes decrease, then sequestration is the cause of the accumulation problem. There is no other scenario that can produce atmospheric CO_2 accumulation. However, atmospheric residence time is a consideration to both scenarios, so the study incorporated a large sample size. It looked at 1971-2022 and sampled data as far back as 1700 to lessen residence time's influence on the outcome.

Using NOAA ppm atmospheric data again from figure 3-S1, I probably don't have to tell you what the study found. But me thinks I will. Falling sequestration volume is confirmed when measuring the year-to-year growth cycle atmospheric CO_2 outflows. Again, confirming something other than emission inflows were causing CO_2 driven climate change. These inflow and outflow graphs are available on the web site. The results also pointed to another fault in our stars, that we've touched on and could use some further explanation.

KEEP IN MIND VOLUME AND CAPACITY ARE SIMILAR BUT DIFFERENT.

Today's CO_2 fast-cycle sink capacities or capabilities are also failing. I'm now referencing how much CO_2 they can take in and how quickly. Essentially, the study is defining the difference from a mature tree or forest to an immature one that lacks capacity. This becomes more evident in higher resolution graphs with longer durations in months and years. Again, the technical writings on the website do supply these graphs. In month to month and year over year (if you're interested in seeing serious downward carnage).

Figures 3, S1, and 4 do provide some basics. They do show the trend. That trend is terrible, alarming, and worse. Capacity continues to decline as global sink volumes decline before and get worse each year after 1850ish.

The sinks are shrinking over time because of clear-cutting younger and smaller trees every year. We started effecting volume eons ago by converting 38% of available global forests into farms, paved streets, and parking lots. And because those sinks are both declining in volume and capacity, the upward atmospheric CO_2 residence time trend grows longer yearly. And that should be expected, knowing that global sinks are impeded by millions of percent and good sinks exist in only 3-5% of the global forests as old growth.

I have the math in a demonstration on old growth sequestration volume coming up. It tells more of this Greek tragedy turned into a contemporary soap opera. But right now, I'm compelled to tie some stuff together. So again, we take a step back to go forward. The topic from earlier, "What didn't get us here, emissions." Let's get that resolved once and for all, if it isn't already.

LET'S TUNE-UP EMISSIONS, PERMANENTLY!

Emissions are a problem, but they aren't the problem humans all thought they were. Let's try this: humans were wrong about emissions in relation to curing climate change but not wrong about the damage they're creating. Reducing CO_2 emissions is not as crucial as increasing sequestration for curing climate change. Don't get me wrong, human CO_2 emissions are bad and harmful, and did I mention bad? They, of course, need to be eliminated. Just not in the extensive ways in which humans currently try to eliminate them. We now know better.

- Earth's natural CO_2 emissions are thought to be around 400-750 giga-tonnes annually.

That amount depends on your source. 575 giga-tonnes is probably a reasonable average estimate, but I'm throwing a hand grenade here, not a dart. The point is any number deemed valid within that correlation with human emissions is compelling to CMS findings and their proofs. The math is coming! Yea!

- Human emissions are documented to be around 35-45 giga-tonnes annually (also depending on your source).

It's time for the mother of all climate bombs. Built out of combining both emissions categories. You can skip the math and proceed to the summary if you like, but I suggest not skipping the green conclusion at the end. The summary of calculations.

ESTIMATING EMISSIONS' EFFECT ON CLIMATE

- Let's say 400 giga-tonnes of natural CO_2 emissions annually. I'm going to use the smaller estimate for naturally occurring emissions.
- Let's say 35 giga-tonnes of human CO_2 emissions annually. Also, the lower estimate.
- Let's use 5.1480 E^{18} kg as the mass of the atmosphere. A common estimate.
- The atmosphere's composition is 28.97 g/mol, so the atmosphere consists of $5.1480/.02897 = 177.7$ E^{18} moles. A ppm is therefore $= 177.7$ E^{12} moles.
- One mole of CO_2 has a mass of 44.01 g, so the mass of one ppm of CO_2 is $177.7 \times 44.01 = 7,821$ E^{12} grams, or FINALLY, 7.821 giga-tonnes of CO_2 in one atmospheric ppm.
- At the time of this writing, the current global atmospheric CO_2 ppm is 421 ppm x 7.821, which equates to 3,292 giga-tonnes of CO_2 that has accumulated in Earth's atmosphere. Way too much, and all the reason in the world to reduce emissions but use sequestration to fix it.

- Earth would like to be around 230 CO_2 ppm (at 1,798 giga-tonnes), so 3,292 giga-tonnes - 1,798 giga-tonnes = 1,494 giga-tonnes of CO_2 that are required to be sequestered from Earth's atmosphere, right here and right now.
- In total, 1,494 + 35 + 400 = 1,929 giga-tonnes of CO_2 that must be dealt with to mitigate or cure climate change to 230 ppm in our atmosphere, generically speaking, "within one year." Okay, so I'm dreaming again. 1,929 giga-tonnes is impossible to mitigate within one year. We'd need five Earth's and CMS working in a perfect status for 15 years. But 435-700 giga-tonnes per year is CMS achievable with just our Earth.
- The remainder, as 1,494 giga-tonnes in atmospheric residence will take a while and can't be ignored because it is definitely the "what" in "what" is causing the climate to change.

SUMMARY OF CALCULATIONS, THE FINAL TUNE-UP OF EMISSIONS:

- *Emissions-based climate-curing attempts address 0.18 percent of the cause or accumulation of atmospheric CO_2. Less than one percent, actually, less than one quarter of one percent.*

No wonder 50 years of emission methods failed to fix or even make a dent in climate change. Mathematically, it was impossible. How we ever went down that emissions path has now become an embarrassing question. We, the People, all believed it, so it must have been confirmed, right? Now, the dreaded reality. Everyone, and I mean everyone, assumed that emissions reductions were the scientific cure for climate change. We have now learned their not. Strange and offensive as it is, they could never be scientifically confirmed with any comparisons, observations, or repeatable statistical analogies. Turns out, it was total guesswork, kind of. But again, we didn't know so it's not our fault we failed. And I'm going to repeat myself here only because I think it's a critical convergence. I also expand the point a little. Until CMS, emission reduction was the best we could have done or tried, so it was good guesswork but now dated by CMS knowledge of climate change being an environmental condition based on sequestration decline.

Anyway, here are the numbers applied: 35 giga-tonnes of CO_2 reduced (that's an at best reduction, a dream) / 1,929 giga-tonnes CO_2 that must be dealt with = 0.18 percent. So now the contrast and compare. Complete Mitigation Science sequestration approach addresses 100% of all CO_2 regardless of its source. Which is why we named it "Complete Mitigation Science."

🌿 *It's time to cure climate change now that humans know what it is. Yes, yes, it is, and I hope before it becomes too late!*

Nice thought isn't it; it rings so true it is not difficult to hear even from a distance. But you'd think I killed a president if you saw the initial reactions received. I was even asked to produce another proof of that, so I did. I took a more extreme approach.

The results of an alternate way of calculating the above proof. Those results are emissions addressing 8% of climate change and sequestration 92%. So, there is that secondary proof as well. I think either proof works fine for CMS credibility. It's time to pop the question.

Having read and seen the considerable difference in human and natural CO_2 emissions volumes and the ginormous difference in the two approaches to curing climate change, let me ask you a serious question,

🌿 "What's more important to fix climate change, emissions reductions or fixing the fast cycle CO_2 sinks that forestry naturally provides?"

You can safely bet that reducing emissions is a great and necessary thing to do. I'd never argue otherwise. Doing so lowers Human impact on our biome which is always good.

However, CMS mitigation is way better at actually curing climate change and mitigating our unavoidable impact to our biome. Note the "unavoidable" part, I need to ask for your patience before I can get to that. Meanwhile, what if we combine both approaches? Now, isn't that an interesting picture painted with triumph's colors?

That wasn't in my planning nor my intention when I learned just how different the two approaches are. However, no plan survives contact with the enemy, the enemy being me again. Consequently, after some deliberation and testing, the study found combining actually works better than my assessment indicated. I was wrong. Emissions and sequestration should balance in the climate-cure equation why not share the mitigation potentials. Which indicates adding CMS sequestration mitigation to any emission attempts ends up working really well. Here's why. Emission attempts are welcome to steal from sequestrations achievements. It can afford it and the math agrees. And emission reductions are good things, but they need a push in the right direction.

Emission-based mitigation efforts actually can't cure climate change without a CMS mitigation offset involved. Because emission things all leak CO_2 in their manufacture or use. Period, drop the mike, Elvis has left the building. For the record, emission reductions cannot cure climate change or even make a noticeable dent.

Now, I've implied the benefits from CMS forestry stewardship can be applied to almost anything. Because they can. So, fixing broken carbon-leaking mitigation attempts? Yeah, CMS can do that with the CO_2 offsets CMS mitigation produces. But what's broken about emission attempts? Well, other than being emissions-based, you mean? So first off, they can never stop climate change because they attempt to treat climate changes atmospheric symptoms instead of the forestry disease that directly causes it. And here's something to really think about that is much pointier, current human perception, prepare to be offended:

- If human emissions all magically disappeared today, climate change would still exist next year and would still be getting worse the following year. You could weakly argue it wouldn't be getting worse as fast as it is now, but it would still be the enormous problem it is for today's generations. Because climate change is created by an environmental condition humans made and not by emissions. Emissions are a variable whereas global forest maturity is a solution.

Again, natural emissions are at 400-750 giga-tonnes to humans' paltry 35-45 giga-tonnes. That natural part isn't within human control, so it will never "leave" nor can humans reduce it. Plus, it's a fantasy to believe we'll reduce 35 giga-tonnes of human emissions to even half of that. We wasted 50 years and thus far have only increased human emissions. That is due to human emissions being part of human domestication. Humans are *emissions dependent*. Therefore, as the song sings, "That's Life!"

I FEEL OBLIGED TO TOUCH ON THE FOLLOWING FINDINGS, AGAIN UNAVOIDABLE.

I might be repeating myself a little, sorry if I do but I needed a lead in.

The problem driving CO_2 climate change is the lack of, and the impeded sequestration in forestry. Not enough sequestration exists anymore to remove enough CO_2 from Earth's atmosphere. SO, CO_2 has no other place to go. But the study showed it used to and can again by increasing global forest maturity. The finding is that CO_2 sequestration is, by far, more critical to climate change than reducing emissions can ever be.

And for the record, emission reductions have many "underrepresented" reasons to remain a priority of human domestication. Reasons include air pollution, environmental poisoning, and human health and well-being in general. So long as they address those bad parts first, the solutions they provide become good for us. The other reason is emissions are completely "unavoidable." We must deal with them in the time and place they occur.

All emissions-based solutions are topped off with what CMS standards call "CO_2 leakage points." Those are things like the CO_2 released while making, using, or taking care of most everything listed in the front of this book and depicted as "Earth's To Do List" and then lined out to indicate we all have, "been there, done that, it didn't work, what's next?" Remember, everything human releases CO_2, even breathing and all human domestication efforts require CO_2 emissions. There's just no way to completely stop CO_2 emissions if humans are going to keep Earth as our address.

Don't get me wrong, those electric or hybrid vehicles, solar panels, wind turbines, and other emissions-based reduction attempts can lower your CO_2 emissions. As in lower your CO_2 footprint. So sincerely from Earth, wonderful job! PLEASE KEEP IT UP! Those things can slow climate change and improve our biome as I described earlier. The problem is that you have to use it for 15-20 years to benefit from a true emissions reduction, and unfortunately, they can never ever claim a climate changing cure. Soooo, why don't you just buy a CMS offset instead and actually do something that fixes climate change? Good question, but I'm sort of biased to be the one asking. Anyhow, that emission reduction stuff problem expands because usually none of those things last long enough. But they can become immediate climate benefits when CMS is included.

Those attempts do nothing to reverse climate-changing conditions, but don't blame yourself. You didn't know. Hey, you're not alone, I paid for all those things too and believed I was doing the right thing to fight climate change. Given the information at the time we were. And in three more years, the expensive car I bought 12 years ago will do as advertised, except now it needs batteries. The attenuation in those solar panels bought 16 years ago makes them junk, taking up space but able to be upgraded and start the cycle again, but they'll be better this time around. Right? Please don't despair with my own regrets of paying big money for these things, although your welcome to because misery does love company. Seriously, humans have a teacher's note for those tests: humans didn't know CMS yet, but now we do! We lived and we learned.

To quote "Peace Sells" by Megadeth, "If there is a new way/I'll be the first in line/but it better work this time!" CMS is proven to work fine and also provides all those previous emission attempts ways to become excellent tools now and in the future. They'll work exactly as advertised in the future; after CMS sequestration standards are incorporated to mitigate their not-so-good parts. We should because they have some really good parts too.

After CMS involvement, the good parts can do as advertised and benefit the climate. But not before. And just to be clear, they can't ever, without CMS. Their plain ole human consumables before because what they provided in environmental benefit is sometimes barely discernible from other consumables. That hindsight makes my Scottish thrawn come out, but I'm calmed when understanding that I didn't know CMS then, either. Still, it sucks to have been on the losing emission reduction team with all the talented players and attitude! Darn, we should have won!

CMS benefits to our environment also continue long after the emissions-reduction-based thing is worn out, removed from service, junked, or hopefully recycled. With CMS involvement, the trees maturing won't be touched for hundreds of years, if ever, so they'll be there to absorb more climate sins as humans create better and better ways to reduce emissions. Who knows, maybe someone will come up with a more environmentally suitable battery, more efficient generation, or even cleaner energy production, yeah, don't bet on it. We have nuclear energy already but the stigma's and misinformation around that stuff can fill encyclopedias. Still, you should check it off the list of Earth's things to do because it does exist and its getting safer.

Unfortunately, it's time to leave the data pool and sit on the towel because I have some unfortunate news. It would be best if you took it personally. I'm not going to hold back, and this news can be scary if not terrifying to some. That is the only warning I'm going to provide so parents you might think about who can read from here on.

While I reluctantly write this section, please keep in mind a couple of things: **First**, my rule to share facts is regardless of circumstances, you have a right to know just how the study says it. **Second**, others and I are working every day towards fixing it all, if there is any comfort in that statement, I could not tell you. **Finally,** the answer is still maybe, and maybe for more than just this particular challenge.

MAKE BAD NEWS WORSE? YES, I'M GOING TO DO THAT EVEN THOUGH I DON'T WANT TO.

I'm not going to sugarcoat this. It's important not to at this point. I mean, you're still here reading, so maybe, just maybe, I'm being taken seriously. Thank you for that. And if you're not, see you in a couple of months! And please remember, the serious stuff I share is not without the hope and real possibilities that CMS stewardship can provide. But candidly, pulling this off without your help becomes a very real, "maybe." Now, I must explain something I really don't want to.

UNFORTUNATELY, THE STUDIES SEQUESTRATION UNDERSTANDINGS ALSO PROVE HUMANS ARE RUNNING OUT OF TIME MUCH FASTER THAN THE UNITED NATIONS CLIMATE PROJECT PREDICTS. I DID SAY MUCH FASTER I MEAN A LOT FASTER.

Am I saying I know better than those thousands of contributing U.N. science experts? Not exactly. The current problem is that CMS is new. As of today, much of it is still in the pipeline to be published, and that worries me, because we really don't have any time to spare. And academic publishing takes forever, years sometimes. So, one can only hope U.N. scientists see it soon enough and then quickly beat any perception problems they might encounter. Also, this is where your involvement comes in: please spread the word about CMS, and quickly! This is why...

- 🍃 The study's unintentional result is that it seriously amplifies all previous climate change projections prior to the study defining climate change. So, modified older study's that incorporate CMS's improved definitions and knowledge quickly become even more concerning with much louder alarms sounding.
- 🍃 The CMS study makes it easier to calculate, model, or make climate projections because the study defined *sequestration impedance and dependence* among other key factors. So, the study removes much of the guesswork from previous non-sequestration-based studies by providing an "almost" empirical measurement and datum(s) to project to and from. I say "almost" because although it is accurate in assessments it is not perfect in its precision, yet. We need that supercomputer to do that. Therefore, the study hits the bullseye every time and only differs in where in the bullseye the study hits.

Here's what we currently know...Around 1950, humans entered into what's called a runaway greenhouse gas effect. It's the most significant problem human existence has ever faced. That is no exaggeration either. Even the Ozone problem pales in comparison. As the name implies, it's a runaway. The greenhouse gas running away is CO_2. CO_2 began running away as parts per million within our atmosphere, starting then and continuing to accelerate even faster now. Looking at the previous Figures 3, S1 and 4 at 1950 you can see what the study implies; the ppm measurement line turned into a rocket ship. The reason why is pretty straight forward. *Demand Driven Forestry*.

From 1850 to 1950, what was left of old growth within most of the Americas was being logged while some of the Amazonian basin also became more actively involved. Land use in the Amazon and America's was also shifting, with people, companies, and governments converting it's forests to farms. Remember humans spent 38% of the global forest available measured from 7000 years ago to today. Surprisingly, the opposite, farms being converted back into forests, was occurring in the U.S. as trains, and then cars and trucks replaced horses. But that conversion lacked any significance in pre-established climate change conditions. We spent to much and shrank our remaining sinks too far.

The CO_2 sink losses by 1950 were immense. Old growth logging had finally worn down the sequestration brakes keeping CO_2 from limitless accumulating within Earth's atmosphere. By 1950, much of the naturally occurring and human made CO_2 had nowhere to go but stack into atmospheric residence. That CO_2 could now stay in the atmosphere for decades, centuries, or in the worst case, permanently. That residence condition had worsened each year even before the 1850 datum. But, by 1950, CO_2 had really been given free rein to become the squatter it is. Ever since 1950 it's been ramping up at alarming rates to any intelligent person's observation. And here's the terrifying thing to me. That buildup started because it accelerated not by decimals, but by mathematically doubling. The problem is compounding and quickening as it does. That's not good and a sure sign our problem is running away while we watch, helplessly, maybe not.

TODAY'S 2-DEGREE WARMER IS TOMORROW'S 10-DEGREE WARMER, AND IT WON'T STOP THERE. IT CAN'T. IT'S A RUNAWAY (WITHOUT CMS'S SEQUESTRATION CURBS).

I'm going to need the blanket you slept under to explain this. If you look at your blanket in comparison to the room you slept in, that blanket took up hardly any of the cubic volume of the room. Especially if we put it into one of those vacuum bags that reduce storage space. As a percentage of the room, it's elements are likely .01% of the room's volume or even much less. But it sure did an excellent job at keeping you warm and cozy!

Atmospheric CO_2 is Earth's blanket and surprisingly the percentage of atmosphere CO_2 is roughly the same as the blanket you slept under when Earth's CO_2 level is at 230 ppm. So, at today's 430 ppm it's not a blanket anymore it's almost two blankets. When it hits three blankets, it's all over but the die off. They'll be no way to stop the runaway as "too late" becomes the description of our fabricated climate failure. I'll explain further in a bit.

The blanket scenario describes the global temperature increase embraced by humans around the study's 1850 datum. It took about 140 years prior to 1850 to achieve the first measurable temperature increase. The next temperature increase, a doubling, only needed around 100 years to appear, and had by 1950. The next increase, another doubling, only took around 50 years and happened by 2000 (maybe even in the 1990s). 25 years later, there's been another doubling in average temperature, arguably all totaling just below +1.5 to 2 degrees since 1850. All because CO_2 has no other place to go other than making more blankets.

But wait! I did say "arguably" because it depends on whose study you read. The doubling effect is consistent throughout all studies, so that is factual, but the average temperature increase is not. The actual temperatures are as inconsistent as a Google search these days, but they all agree with doubling as they rise. So, because of the extended sample size of science papers on the subject, CMS concurs with the bases and the effects they established. They're doubling and taking half the time to do it. Plus, the average global temperature is higher. It's hard to tell how much higher, but all the reports I used all agree on that as well, too high and getting worse fast!

Which is going to boil us all down into real trouble, I'm trying to lighten up this news. As more CO_2 accumulates in the atmosphere, the runaway effect will only come quicker and quicker with higher average temperatures. And it won't stop without CMS intervention.

So, guess what, we really are in a lot more trouble than any political or current scientific agenda might bravely or honestly address, today. Why? Because they either haven't or are just now seeing the CMS study come out. And perception of new data, yeah that thing, is always difficult. Argument will ensue, I'm sure. However, facts are what they are. They can spin them or look for insignificant points to soap box couching but ultimately, they'll lose. Getting them to lose before it is too late is a fact driven challenge.

So, climate change is a lot worse than previously understood by emissions reduction science. Is there any hope? Yes, there's hope, but it's in short supply. Do we have the time to fix it? I can only answer again with a maybe. We must stick with facts and drive the points home if there is to be any real hope.

Remember, sequestration is more critical than emissions, and accepting that fact only helps to fix our runaway problem. I also need to again mention that hope and time are hindered by the first climate-changing problem: perception. Yours. So, the next section demonstrates why yours is also a problem to contend with facts. Time required to convince is precious and almost pricelist at this point.

A SIDE TRIP AS AN EXAMPLE TO A MAYBE.

I drive down from Oregon to visit family in Northern California frequently. It's a 6-hour drive. We've been doing that since moving from Texas to Oregon many years ago (Hook'Em!). I remember the first trip vividly. Snowcapped mountains, Mount Shasta with its glaciers looming, everything covered with immature trees. To my naive eyes it was a peaceful drive, green no matter the season, and the air smelled fresh. It was refreshing, vibrant, and there wasn't much traffic or road craziness until you arrived in Sacramento.

Today, that drive is everything but enjoyable. The land in Oregon and above Redding, CA along Hwy. 97 looks like the burnt end of a used wooden matchstick from several large fires that erupted in differing years. It smells of charcoal and is as hot as a desert tortoise's back at noon in the summer. There are no snowcapped mountains unless it's late winter, and Mount Shasta's glaciers are all but gone or hardly noticeable. It only took sixteen years for all that land to morph into its climate-changed form of today. And it's getting worse.

Climate change's temperature increases greatly influenced my drive's view with forest fires and less snow. More specifically, the fires are greatly influenced by what causes and is perpetuating climate change today, now more than ever. *Unconstrained and constrained deforestation* makes those fires increasingly possible as increasing global temperatures increase fire season durations and suck the moisture out of forests. But that really doesn't define the impact well enough.

I've explained the impact of *demand driven forestry* and how *constrained deforestation* is the primary result. But something else is killing forests due to climate change—bugs.

Changing our biome, as we have, make the biome more conducive for some species and less to others. In this case it's wood boring beetle's. And that's not good for entire ecosystems, like forests. Entire forests are being wiped out overnight.

Beetles are thriving simply because the average temperature has risen enough for some beetles to expand their range into higher and usually colder elevations. These beetles are really hard on trees and one answer to why I answer maybe it is too late. They bore into a tree and lay eggs where they shouldn't. The eggs, nests, and wood eating offspring destroy the tree's vascular system and thus the tree's ability to live. The result is the beetle infestation eventually kills the tree then the forest. The beetles don't mind killing it because the dead tree becomes even better habitat for them to spread to alive trees the next season and start the process again. It's a real problem that in turn helps fire consume vast amounts of forest, among other reasons just as nefarious. Like immature trees overcrowding forests and entire ranges in regeneration at the same time that makes a bunch of wooden matches for an idiot, careless adult, or lightning to ignite.

CMS must compensate for what we have to work with in order to restore it back to what's needed. Except this time our geo-engineering won't ignore CO_2 sequestration and will restore the required symbiotic relationship. It isn't a given, but

I hope we can do it quickly. To anyone who states otherwise I say, "let's get ready to rumble!"

An important part of the study's mitigation plan is to go down fighting if Mother Nature and humans do not cooperate, but she will so we will. Okay, neither she nor we have ever before. With laws, we have and did for the ozone layer. But she has never cooperated. In fact, I'm quite sure she's okay with us continuing to fall on our sword. I'd even say she wants us to. I say that because she darn sure put a time limit on us fixing our previously incorrect forestry stewardship. And now is that time, not tomorrow, because there is no tomorrow as far as she is concerned, she really doesn't care what she looks like in the morning or to anyone so long as she gets her way.

Okay, I think I made my point, but I want to be sure, so... Mother nature isn't going to wait for us to continue to do little to nothing. Her plan goes on without us. So. it's time for us to get it all fixed and done quickly, because it's cheaper to win her back than get permanently kicked off the planet.

Speaking of planets, now it's time for some Star Trek! "Space the final frontier."

SCOTTY, I NEED WARP DRIVE NOW!

Star Trek's Scotty character is an engineer, so his character likes math as much as I do. They tell me not everybody likes math, which goes against my grain but to the "majority" their own. For the lack of math in this book, you can thank my wife Toni, Theo, Claire, Dave, and Gene Roddenberry, Star Trek's creator and pretty much the entire world's population in general. Mostly Gene. He never let Scotty explain the math in Star Trek because he knew nobody wanted to hear or see it. Plus, they made it all up. Instead, Kirk always shortened the conversation by exclaiming, "Scotty, I need warp drive, now!" And all the calculations to repair the damaged warp core were done, whoosh! So instead of math, how about a little something else to fill the gap?

A few things to check off before moving on.

1. Every year a tree is allowed to mature, it takes in more CO_2 from the atmosphere then it's previous year. It does so in an accelerated manner by applying a 3-8% per-year growth rate. Exciting, right? A note on this. If you look around for data on this, be careful. Many foresters say tree growth slows down after so many years. It does, but I'll take the 3% growth of a 20,000

lb. mature mass over 8% growth of a 200 lb. immature mass any day. Yes, forest maturity improves sequestration, period.

2. The really sticky part to fixing climate change is Earth naturally emits 400-750 giga-tonnes of CO_2 yearly, depending on the report you read. 400 is the lowest number I've found, and I have difficulty buying just 400. It seems too small. But I genuinely don't know, nobody really does. It's unmeasurable but it is estimable and is therefore factual. Why? Even though we lack precision from an exact number we are accurate in assessment that it does exist and it's an exceedingly high volume.

3. Humans globally emit 35-45 giga-tonnes of CO_2 yearly. Again, it depends on the report you read. And there are hundreds of reports. I use 35 giga-tonnes in my estimates, which I think is a solid estimate based on the law of averages and the rule of large numbers used in statistics.

4. *Constrained deforestation* is widespread and affects 85-95% of global forests. It's closer to 95% than 85%. The 85% comes from the bottom of the scale as a statistically applied amount. Again, it is more likely

closer to the 95% which only make the study's findings far more significant in outcome but less conservative in approach. Sequestration findings applied to either range tells us that climate change is worse than we thought. Which also makes it impossible to argue against the study's outcome. What I mean is demonstrating the impeded sequestration finding at 85% is made much easier to demonstrate at 95%.

5. *Constrained and unconstrained deforestation* continues to increase with time. There are significantly fewer mature trees each year because smaller and smaller trees are harvested. That also creates *tree degradation*. Plus, replanted trees and tree farms are expanding to *demand driven forestry*. That is both good and bad. More trees is good but more immature trees never allowed to mature is unbelievably bad. The expanded replanted land also creates more *carbon humps* and more land expelling CO_2 and methane emissions during the hump.

6. The *carbon hump* reality is because replanted trees can't absorb the CO_2 released by the clear-cut for 20-30 years. Clear-cut, naturally regenerating, and newly replanted forests emit massive amounts of CO_2 and

make so called renewable wood products CO_2 emitters and not as they are advertised as "green" renewable products. Only 20% of the biomass logged is considered carbon neutral. The rest of what the tree had stored is back in the atmosphere within 10-20 years.

7. CMS demonstrated that global forestry fast-cycle CO_2 sequestration is not running out: it's already way out, gone, diminished, absent, lacking, and by millions of percent. MILLIONS OF PERCENT. The study classifies this effect as *Impeded Fast-Cycle sinks*.

8. Out of the entirely of all global forest, only 3-5% of mature forests remain. The math states the remaining 3-5% of old growth is responsible for 85-90% of the miniscule volume of CO_2 currently coming out of Earth's atmosphere today. The outflow volume must increase. The other 95-97% of global forestry is too immature to do any good and likely stuck in a *carbon hump*. It is; however, exactly why CO_2-driven climate change has the opportunity to make all animals extinct and exactly what caused CO_2 driven climate change.

9. The 3-5% global old growth is the primary terrestrial net CO_2 negative sequestration source on our planet. Nothing compares to its ability in CO_2 sequestration

and sequestering carbon. No batteries, plugins, energy, software, hardware, employees, or internet connection is required for photosynthesis to work, its solar powered and not manufactured by humans. That provides CMS mitigation all the chances it needs to fix climate change, except money to do it.

10. Finally, it's estimated that 2 trillion trees are on Earth. Which is a good thing, sort of. Russia has over a trillion by itself. Globally, trees grow on around 10.3 billion acres as forested land. But roughly 85-95%, or up to 950,000,000,000 billion acres are regularly logged when they're typically under 40 years old and today's reforested trees average less than 40 years old. Yikes! And remember what I mentioned earlier about less acreage is needed with more mature trees. Which means, we have trillions of immature trees because we don't have mature trees taking up more space with much higher sequestration rates. Yikes again, I know.

11. 38% of the global forest has disappeared to human expansion over the last 7,000 years. The majority of those forests disappeared in the last 200 years.

12. Emissions are unavoidable and take two forms. Natural and human. Human domestication requires

emissions least we kill off 75% of the human population and go back to nomadic life, because Earth can only support about 25% of today's population living nomadically.

13. CMS fixes climate and there is no other way possible. Plus, CMS has many residual benefits.

TIME FOR APPLICATION OF THE STUDY MADE PRACTICAL WITH THE DREADED MATH!

So, let's eliminate just human emissions with trees. We'll ignore natural emissions initially. So, knowing what we know now, it really isn't that difficult, or is it? That depends on if your still with emission's or converted to sequestration's proven results.

- Yearly, human emissions are roughly (35 giga-tonnes) / (divided by) (CO_2 giga-tonnes per acre.)
- Because humans need to remove giga-tonnes, let's see how many lbs. are in one giga-tonne. First, 2,204.6 lbs. are in a metric tonne. "Giga" means one billion (1,000,000,000), so a giga-tonne is one billion tonnes. So, 2,204.6 lbs. per tonne x one billion equals 2.2046 E^{12} lbs. in one giga-tonne. Say what? Scientific notation is tricky, so let's convert that to decimal to make

it easier to see. So, 2.2046 E^{12} is the same as 2.2046 x 10^{12} and equals 2,204,600,000,000 (two trillion two hundred and four billion six hundred million) lbs. per giga-tonne. That is a heck of a lot of CO_2's!

- So, 35 Giga-tonnes CO_2 yearly x 2.2046 E^{12} lbs. of CO_2 = 7.7161 E^{13} lbs. of CO_2 needs sequestered.
- Here's what *constrained deforestation* is currently providing Earth with its estimated and paltry 1,100 lbs. of CO_2 sequestered per acre.
- As a reminder, 8.5 to 9.5 billion acres of forest (out of 10-10.3 billion global forest acres) are unprotected and within *constrained deforestation*. 8.5 billion acres x 1,100 lbs. of CO_2 sequestered per acre = 9.35 E^{12} globally sequestered lbs. of CO_2 in one year, stated as 9,350,000,000,000 (nine trillion three hundred and fifty billion lbs. of CO_2 total). And now lbs. back into giga-tonnes. 9.35 E^{12} / 2.2046 E^{12} = an estimated and paltrier 4.24 giga-tonnes of CO_2 is being sequestered under today's *constrained deforestation* using the Earth's unprotected and logged forestry.
- In comparison, the last remaining mature forests globally occupy around 1.8 billion protected acres, roughly 17% of the total forest area of 10.3 billion

acres (depending on the source; I'm using United Nations data which I think is too high in acres, 17% is thought as "overstated"). For the record, these trees are off the sequestration charts. Because they're so mature, 300-plus years old, they can sequester tonnes per tree yearly. Again, being mature does not automatically mean they can. Species and location also play an influential role. However, marketable tree species, well, they pretty much all can with enough maturity. Especially firs and evergreens.

- I estimated they could be sequestering around $2.032 \, E^{14}$ lbs. of CO_2 annually, as in 203,200,000,000,000 trillion lbs., or 92 giga-tonnes, of CO_2 yearly.
- The study measured atmospheric ppm inflows and outflows, but much of what was not measured in the study never made it to the measurement stick before these super-sequestering trees got it! And we need a lot more of that!
- The two examples combined equal 96.24 giga-tonnes of CO_2 per year sequestered. Not nearly enough, even when combined with other sequestration sources like the ocean, to sequester both natural and human emissions, again, around 435 giga-tonnes. Hmm, I wonder why CO_2 is accumulating

in our atmosphere? Oh, it's because it has no other place to go.

- What is significant within these examples? They prove a point using both CMS data and an alternate data source.
  - 5-15% of the global forest is performing up to 90% of Earth's current CO_2 sequestration annually. So, any improvement in the other 85-95% of the global forest is a winner-winner chicken-dinner for curing climate change. And why the CMS study can provide a solution.
- So, at age 59, the ability of the tree in the earlier appearing tree maturity example expands to approximately 134.1 lbs. of carbon sequestered from 492 lbs. of atmospheric CO_2 that growth year. I dropped the number to 109 trees per acre because they're older and need more space.
  - At 8.5 billion acres x 53,628 lbs. of CO_2 per acre is 206 giga-tonnes of CO_2 sequestered from the atmosphere by increasing the global average tree maturity to 59 years in the unprotected forest. A 47% increase was achieved in only 9 years. That is a wow! Add 206 giga-tonnes to 92 giga-tonnes of existing mature forestry, and what do you know! 298 giga-tonnes of CO_2 are sequestered into forestry. That should give you a warm fuzzy feeling about fixing the climate change mess!
- Now apply an average 6% annual tree growth rate using the 8.5 billion acres. I'm also dropping to 79

trees per acre to make some room for the bigger trees, which in reality isn't really needed, but I like to stay conservative with numbers, I'm also using a more conservative tree species because I want you to see maturities effect with a lest performing tree species in the study:
- At age 66, 225 giga-tonnes for a total of 317 giga-tonnes CO_2 are sequestered per year.
- Age 71, 301 giga-tonnes for a total of 393 giga-tonnes CO_2 sequestered per year.
- Age 78, 453 giga-tonnes for a total of 545 giga-tonnes CO_2 sequestered per year.
- Age 84, 643 giga-tonnes for a total of 735 giga-tonnes CO_2 sequestered per year.

And it all continues to improve with age. And why I don't care exactly what number natural emission giga-tonnes are at (400-750). CMS doesn't care where emissions come from, it only cares about where trees put them, permanently. Well, for around 800 years at least.

And that brings me to the point of contention that I insisted was made in this book. Even if CMS is fully executed for a decade or two, humans will still be in greater jeopardy from a much more natural and recurring problem interacting with our previous climate sins. Yes, I hate to say it, but the study turned up more scary stuff that provides another reason to only answer "maybe."

IT'S HAPPENED BADLY BEFORE, STOP HELPING IT NOW

Kaboom! In 536 CE, 1,487 years before this book, the Krakatoa volcano erupted and changed humanity forever. That eruption created a climate chaos that set human domestication back centuries. Many theories speculate it may not have been one volcano but a combination of eruptions. More recently, scientists who study tree rings and a geologist came together to isolate Krakatoa as the main culprit. Whatever happened, humans do know the effect. It was historically recorded on a global basis meaning there is 1st hand and global experiences recorded in abundance.

Some think this event made modern humans behave the way we do. Maybe it hit society's reset button. Many historians, even the majority, believe it did. What makes it significant to the study or maybe even worse? Is, it wasn't the only time this sort of thing has happened. Which makes me concerned. So, the following scenario is more concerning as an extinction level event than a huge asteroid hitting Earth is.

Why?

Well, we actually have technological capability that can deal with a rogue asteroid but there isn't a thing we can do about a volcano erupting, yet. Except to be prepared for it to happen and that is where our climate change problem of today is not helping us, at all.

THE DARK AGES ARRIVED IN ONE WEEK

After the eruption, a colossal tsunami killed many, but that was nothing like the numbers that died from the volcanic dust cloud or the CO_2 the explosion(s) produced. Those environmental effects were far more costly to humans than its 40–60-meter Pacific Ocean sloshing.

For the record, today, all volcanoes combined don't outproduce humans in CO_2 emissions in one year, normally and in most years historically, there are many exceptions. Conversely, massive eruptions like the one in 536 CE emit enough CO_2 and other discharges to change the global climate, and they do it all at once, in days or weeks not years.

But CO_2 increases in the atmosphere are not only because of the eruptions. Sure, the eruption emissions can, of course, be gigantic. Other reasons for the increased CO_2 in the atmosphere, you could say, are not expected and just not easily relatable to daily thoughts.

Thanks to studies on tree rings and historical accounts from across the globe, we know the world went through 1.5-3 years of a mini-ice age (dependent on where on the globe you lived; the further north or south from the equator, the longer the mini-ice age). Trees living during that time grew little, if at all, and many died thanks to the suspended dust in the atmosphere that blackened out the sun for years. Crops didn't fail, they never even sprouted. Little to nothing grew. "Everything only withered and then died of the ice or the dark." That's a quote from a French Monk during the global crisis. There is also Chinese and other historians in that loop. What makes those documentations significant is they clearly indicate sequestration was also suspended, so little to zero photosynthesis occurred during those 1.5-3 years of mini-ice age. And guess what, those giga-tonnes of natural emissions and a little human emission to keep from freezing stacked up in the atmosphere or froze to the ground for the thaw to occur. And it did.

Massive amounts of CO_2 were also released during the eruption(s). But that single CO_2 burp was compounded with another massive CO_2 release, and here's the gross part: the mass dying of plants and animals (including humans) also added to atmospheric CO_2. And without sequestration working well or really not at all, all the CO₂ dogpiled into atmospheric residence conditions. Sound familiar? Yeah, like todays runaway effect maybe? Oh, and don't worry, this all gets worse. Time to pay the piper arrived.

By 538-539 AD, some of our predecessors looked at the sun for the first time in 3 years. They knew they had survived by being murderous thugs working for a ruthless pope, cleric, warlord, dictator, or king who had hoarded food like gold and fed only the loyalists doing their bidding. We're all descendants of people who survived. Yikes!

By the end of the mini-ice age, nobody had been living well, if at all. Just enough of everything survived the viciously cold and blackened days to make a decent try at starting over. "The Dark Ages" really defines it well. All of human civilization collapsed, vanished, or hibernated for hundreds of years. That's like saying all government disappeared because the people in government were too busy trying to find food for themselves. Total anarchy. Truly little of anything lived through it, and that's because when the mini-ice age ended, everyone's troubles only warmed up with the appearance of the sun.

When the mini-ice age stopped hiding the sun, it was followed by a short timeline of massive floods, then 10 years of catastrophic global drought in the Northern Hemisphere and approximately 20 in the equatorial regions. Those droughts alone may have killed more than those dark days of cold. Imagine this: just as the sun comes out and you think it's going to be all right; your flooded out and don't receive enough rain (after a flood) to grow crops for ten or more years. Ouch!

You might be asking what all the Dark Age stuff has to do with CMS. Well, eruptions have become an even more significant problem today. Here's the understanding we now have. Let's start with the flooding. That was caused by the sun suddenly appearing and thawing everything and all at once. Which put a lot of water vapor and CO_2 into the atmosphere with all that stacked up CO_2 to keep it warm resulting in biblical rain events. And those rain events were massive, off the charts and nothing like anything ever heard of since. The next thing to deal with is drought. That was caused by the excessive CO_2 buildup from the eruption, the die off's and almost zero global sequestration because photosynthesis was impeded or even suspended. So, Earth could not absorb any of it until photosynthesis recovered, eventually, it did recover but took a while to catch-up so it could sequester all the excess CO_2 from the atmosphere allowing the suspended water vapor in the atmosphere to cool and fall as normal precipitation patterns, ending the drought. The reason it took longer to end drought along the equator is the eruption was nearer the equator than Earths pole. So, way more CO_2 existed there than further north or south. The corellas effect of Earth spinning and the geographic size (as a much bigger diameter) makes the equator area more prone to climate change or events like the one in 436 CE. Now to explain today's volcanic issue.

The 536 CE event happened when the global fast-cycle sinks were reasonably healthy, and the atmosphere had low CO_2 ppm levels. What would happen today? Do today's highly impeded fast-cycle sinks make you feel safe? How does the 430 ppm already stacked in our atmosphere make you feel? I'm concerned enough to write about it. Are you okay knowing about it? I did say I was going to make this worse, so it isn't over yet.

Add this in. Krakatoa had a minor eruption in 1863, and most research indicates it added to the following decade of CO_2 ppm in the atmosphere with a few years of global drought after 1863. How much? Unfortunately, that's highly debatable and one of the reasons I chose the earlier date of 1850 as the CMS datum. Yes, CO_2 spiked up, but not by particularly concerning numbers. It was too small of an eruption. Even though other adjacent eruptions added to Krakatoa's, the effects were nothing like 536 CE. In retrospect, volcanic activity from the late 1700s to the 1880s was tremendous, much more than we've seen in the last 100 years. So, here's my take as far as eruptions go.

Eruptions are rated like earthquakes using the Richter scale. The 536 CE eruption is currently being researched. It's a more historical accounting and its effect is not as well documented statistically as it is historically. It's thought to have been a high VEI 7, because of the well-recorded carnage that followed. But it doesn't rank high enough in size to make the top ten eruptions on Earth. The 1863 Krakatoa eruption is documented more statistically and considered a VEI 5 or low 6. The 536 CE eruption was far more destructive in all comparisons and probably a total of 2 magnitudes of 10 higher in rating. Each rank is a factor of base 10 (log10), so ten times more powerful than the previous magnitude. Oh, and for those who remember Mount Saint Helens, that was tiny compared to what I'm talking about, a VEI 5.

Our global CO_2 fast-cycle sinks are only 3-5% usable today compared to their historic capabilities before the 1850 datum or the 1863 eruption(s) or 536 CE. They were really good in 536 CE but not perfect by any means. There weren't enough humans globally to challenge forestry supply with demand yet. Europe and China were minor exceptions. Those regions had been practicing *constrained deforestation* for thousands of years before 536 CE. However, the Americas and really most of the world's forests were still untouched. And that saved us from a worse fate. To explain further.

1. Atmospheric CO_2 levels are increasing upward at a concerning 422 ppm in 2022 instead of an estimated 280 ppm before 1850. In 536 CE, they could have been even lower. Approximately 230-240 ppm is not out of the realm of possibilities because there weren't as many humans alive in 536 CE.
2. Another 536 CE Krakatoa-like eruption could change the current 422 ppm upward in a massive way—100's more ppm in a concise period, like a week. Closer to the volcano, it could be 1000s upon 1000s of increased ppm.
3. Photosynthesis is seriously hampered or suspended during mini-ice ages, there's no sun. CO_2 therefore stacks in the atmosphere and waits for the sun to reappear and cause serious droughts. Kinda of like todays decades long droughts isn't it?
4. The carnage of a mini-ice age also releases CO_2 as plants and animals starve, succumb, and then rot (for example, when your body loses weight, much of it is

released as CO_2). Yes, humans are full of carbon. That also happen when the sun reappears and all at once during that initial and very quick thawing period.

Okay, knowing all that, let's say Krakatoa, Yellowstone, or Toba, all previous VEI 7-8 eruptions similar to the one in 536 CE, occurred tomorrow. Remember, there are 20 of these big volcanos globally and over 1,500 smaller ones. And nothing says any of those 1,500 can't grow. For the sake of simplicity, one of the 20 could, by itself, quadruple annually released emissions of CO_2, or even a heck of a lot more, in a single event.

Here's my take, and for what it's worth. Historically, giant, or combined eruptions were just a bump in the road of human domestication because of healthy forestry and their mature CO_2 sinks, just like during Krakatoa's 536 CE eruption. Sure, many died, but humans as a species were able to recover. Today, with nowhere for that CO_2 to go, it could accelerate the runaway greenhouse gas effect past the point of any possibility of recovery. There is no prevention or long-term predictions for volcanic eruptions.

The concern specifically is that volcanic activity is both expected and unpredictable. Expected, as it's part of Earth's natural order of operations. Unpredictable, because nobody knows when or where the next eruption will be or how big it'll be. But it will happen, and that certainty is what concerns me.

Because now, we don't have healthy sinks and already have 421 ppm stacked in our atmosphere. The study exploring volcanic activity tells us our negligence of forestry sequestration has placed all animal life susceptible to this particular failure mode. Basically, we're now asking for it.

In case you missed that point let's get a little closer to home. A little more recent in time. And another reason to add to maybe. Yes, I'm making it even worse.

1816, the year without a summer. This was long before the CMS 1850 datum. I've mentioned, sequestration was damaged long before the datum but not as bad as is now. Kaboom again. This climate anomaly occurred in 1816. It was predominantly a volcanic winter event caused by the VEI 6 eruption in 1815 of Mount Tambora, Indonesia. It was possibly worsened by a minor eruption in 1814 of Mount Mayon in the Philippines. Its emissions, including acid droplets, starved the planet of the sun for almost 2 years, and its "red" atmospheric effects from the acid droplets lasted over ten. Although, it did not have the completely blacked out sun of 536 CE. The eruption(s) did diminish the sun's rays for a couple of years, like using red shaded sunglasses. By the way, neither effect on the sun is from CO_2. That effect comes from the ash and dust particles. The CO_2 effect comes after the sun reappears.

At the Church Family of Shakers near New Lebanon, New York, Nicholas Bennet wrote in May 1816, "all was froze," and the hills were "barren like winter." It was snowing, as a blizzard, June 7th, and 8th in the same area.

A Massachusetts historian summed up the disaster like this: "Severe frosts occurred every month; June 7th and 8th, snow fell, and it was so cold that crops were cut down, even freezing the roots... In the early autumn when corn was in the milk it was so thoroughly frozen that it never ripened and was scarcely worth harvesting. Breadstuffs were scarce and prices high and the poorer class of people were often in straits for want of food."

Millions again died globally from starvation; Thomas Jefferson, yes, that Thomas Jefferson, went into debt because of two years of crop failures. It was even worse in the South Pacific and Asia, being closer to the volcanoes and lasting much longer because of that geographical result explained earlier. And with that being said, the quicker CMS is put to work for us, the better. Because this sized geological bump in the road is about all it would take to make animal recovery questionable with todays 421 ppm level. It is not a matter of if it will happen again, it is a matter of when it happens again. And maybe.

CRASH! IT'S NOTHING, BUT I CAN FIX IT

Climate change is a global problem. It is everyone's problem. It is a human problem. It is an animal, plant, and mineral problem. Whether you're a newborn or not long for this world, it's your problem. It is not what humans want to deal with, it's what humans must deal with. And most importantly, it is a geo-engineered problem that CMS can reverse. Undesired terraforming, you could call it that. The study calls it an environmental impact and has little to do with human emissions. Which is actually good news because that makes it entirely fixable. But not in the way you have been told by the media for the last 50 years. Those years have cost us. Now, we are all left hoping we can fix it soon enough.

SO FAR, NO GOOD.

Time to paint it black. So, what are we in for by ignoring climate change and continuing to try and resolve it with politics and nonreproducible, unproven science? We have the perfect example. Have a look at our sister planet Venus. Do you know how Venus became Venus? Volcanoes spewed

CO_2 that created an atmosphere so thick and heavy with CO_2 it crushed its outer surface layer into one planet sized shell (Venus's atmosphere is about 90 times heavier than Earth's). Due to that, Venus does not have any continental plate drift that forms a magnetic shield like Earth does. Accordingly, it is radiated by the sun, it's being microwaved. Long ago, it had a magnetic shield. So, Venus also had an atmosphere that was remarkably similar if not identical to Earth's. But it all went badly because of natural CO_2 emissions.

Venus suffered a runaway CO_2 greenhouse effect from volcanic activity, that all started small and grew to make Venus the uninhabitable nightmare of today. That CO_2 that ended it all for Venus is the same CO_2 we must contend with now on Earth. Except it took millions of years for it to naturally happen to Venus, but Venus didn't have humans greatly accelerating the problem. Yeah, we are way more efficient at screwing a planet up and a lot faster than natural emissions are.

Earth can quickly become and has been getting geoengineered into becoming the planet, Venus. If humans keep ignoring sequestration it will happen sooner than you think. It will not take millions of years. At our current rates, it won't take much longer to hit a point of no return. Another top 10 volcanic eruption could do it right now. Yeah, it's scary to think about but true, and that's what makes some of the study's results so terrifying especially when previous emission-based climate projections incorporate CMS sequestration. But you might be thinking humans are pretty smart, we'll just adapt! Many politician's and oil companies think so. Maybe even timber companies. Okay, lets apply more black paint and end that dream right now. Because that is impossible, you can't hide or adapt to climate changes effects.

So, you want to talk about surviving climate change, do ya? Okay, so maybe we can get away with it for many more years but generation X won't. Unfortunately for the X's and down, you read this book and now know better, much better. And what we know can be simply stated...There is nowhere to hide from climate changes plans for all animals.

Planetary science provides the doom of a long-term forecast, the short term is actually worse. We can't survive the getting there because people will make it worse. First the long term. Good luck with a survival shelter since a CO_2-saturated atmosphere can weigh 90 times what Earth's does now. That pressure is like being a mile or two below the ocean's surface, yeah, bone smearing not just crushing. The Earth itself would actually become compressed and decrease in size. That's happens after the oceans boil off into the atmosphere, making it even hotter via water vapor. Bring sun block for 200 plus degrees. All the while solid rock miles deep shatters like glass, but rock for a couple of miles down or so will eventually melt together ending the magnetic field that protects us from solar radiation. All of that nightmare occurs just after the runaway greenhouse gas effect kills all plants with excessive CO_2 fertilization, while cooking them with ultra hot temperatures, and biblical storms with 1000 mph winds. The animals died off a long time before that due to a CO_2 atmosphere forming that will last billions of years, so pack enough snacks and drinks for several million generations and bring oxygen bottles. Oh yeah: no more drinkable water anywhere. It'll all be converted into $SO2$ or CO_2 or much worse like hydrogen acids, ammonia, rocks, and you can't drink that stuff. I mean you can, it might be better to die from drinking that stuff then be turned inside out by the atmosphere's weight, microwaved by the sun, turned into jerky by the radiation or cooked like a baked potato while trying to live miles under the surface. There is no hiding from climate change results. To Earth, it's like a light switch, your either on Earth or your off, as in extinct. Oh, and Earth has already had 6 extinction events so its experienced at evictions like the one going on now. But it takes a while to flip that switch if humans are not involved. But humans are involved, aren't we? And now, the short-term shade of black.

Oh, but you might think it can't happen overnight. And yes, you are correct, mostly. I say mostly because the runaway has a 75-year jump on CMS already. In another 30 years we will know more of Mother Nature's wrath than we ever cared to. To make it worse, it looks as if it becomes too late around the halfway mark to 30, and only if we have luck on our side. As I mentioned, the study gave us a way to measure sequestration with CO_2 ppm Delta's. They have gone negative two times in the last 7 years and never ever before. The effects insure the results. What are those effects?

Well, I'm worried about my kids experiencing a further accelerated domino effect because it has already started falling apart. Things that you have already experienced as climate change are getting worse, quickly. Like access to food and supply chains in general, unbelievable heat waves, huge forest fires, rising sea levels, rising water vapor levels, methane increases, sea ice melting, long term droughts continuing, fresh water scarcity, plant and animal extinctions continuing and happening faster, and storms that appear out of nowhere and then intensify so fast and to biblical levels and way out of normal "storm" seasons. All measurable and escalating.

Yes, all of that is happening now and accelerating faster than one might believe. We know the results because all those climate driven events are scientifically observable right here and right now. And they all get worse every growth cycle we ignore sequestration. But when is going to be when you might be wondering?

Three decades four tops, the year 2054 is not a new year we're all going to celebrate, not without sequestration restored. Around then the study projects survival will become a true commodity, and not a given like in today's free world economies. Yeah, it is that bad and now predictable with greatly improved confidence. Just like the frog put into the pot of cold water and then the heat applied; we aren't noticing the pot beginning to boil because it was room temperature when we jumped in. Just have a look at what's happening to country's along the equator now, quickly intensifying storms, floods, famine, bug infestations, disease, mass migrations, wars, etc. Hell, if the world didn't send food to some of the equatorial country's people couldn't live there and many of them are immigrating elsewhere, returning to nomad conditions you could say. Mostly because of drought, CO_2 driven drought.

Wars over resources/economics are just beginning and not just because of the long-term droughts either. Country's go to war now for their own low yields or to acquire better farms and additional economics (Russia). When Russia invaded Ukraine the first thing Russian's stole was grain and the first thing they blew up were factories and mines that competed against them. Evil empires, like the 436CE dark ages, are forming and joining forces these days. Another example, Islamic Jahad, you can't have one of those without guaranteeing food to the martyrs surviving family. Many of those folks know hunger well already, so it's no surprise to me when mothers send their son's to fight "infidels." Food in many of the Islamic country's is used to control populations. Count on more of that everywhere because country's like Syria and Iran can't afford enough food. Plus, military options are less expensive and ultimately create less mouths to feed by increasing mortality rates and making refugees. So, they and others have nothing to lose but people.

Over the last 75 years, what was once forest or grassland is going into desertification in many parts of the world, especially everywhere in the middle east, ah but not Israel. Hmm, wonder why? I'll tell you why, democracy and free markets create and instill environmental consciousness regardless of personal faiths. And now to droughts, the first sign of excessive CO_2 ppm. It's everywhere and getting worse, it's also the third or fourth decade of drought in much of the world. That historically, is far from normal. A year or so of drought is normal, without large volcanic eruptions that is.

- When I'm asked how much time we have my answer is simple, "we don't have any time."
- When I'm asked can we fix it my answer is simple, "yes, CMS can."
- When I'm asked will we fix it my answer is, "probably not." Please, prove me wrong.
- When asked can we fix it in time my answer is not so simple, "maybe."
- When asked what the odds are I answer, "not good."

My plan to address those questions is this, "There's only one direction to travel, I'm going down fighting climate change or not going down at all because we fixed it." In short, "I chose to fight so my Scottish Thrawn wins either way." My family's crest and moto is "I only hope for a fair fight" Truth is, I just care for the chance to fight.

LET US GET TO WORK IN A POSITIVE DIRECTION BECAUSE WE CAN!

We need to mature trees beyond their current harvest ages. That is the entirety of CMS's stewardship goal and how to repair CO_2 driven climate change. With enough time and enough trees, we beat our current climate change failures and our historical climate sins.

That will, of course, require money. But here is the good news. Fixing it all is cheaper than ever. You could even say it's on sale for a limited time only! Any money to CMS goes directly into curing climate change and CMS doesn't require what the world expected to pay. Inexpensive and measurable results! And shouldn't any climate mitigation effort be able to say that like CMS can?

You're darn right they should, but no one else can! They might be able to stretch it and say they're lowering emissions, and yeah, sure they are, but they're not fixing climate change without CMS involved, and now we also know why that's a problem. So, don't buy so quickly into that hype anymore! Not without sequestration someway involved.

Sorry, the Scottish thrawn got out again. Okay, you can believe in the emissions-reduction hype, but tell them you want it paired up with CMS, so that it becomes truly real climate mitigation. Even better, don't wait for them. They will only disappoint. Offset it yourself! Hit the EWC website and buy an CMSco_2t carbon credit. It's easy, and by doing so, you're fixing climate change for real. And for the record.

Maturing trees is not going to be free or easy just because we need them to. It is going to require profoundly serious coins although not anywhere near the cost of those other emissions-based things. Still, humans, forestry, factory's, OPEC, and everyone must buy into CMS stewardship. No one can just step into a globally well-established wood biomass industry and say, "Sorry, you can't earn a living or have profits because it's all your fault." It's not their fault, they are just as human as we all are and didn't know CMS either. Give them time we don't have? No, I didn't say that. They won't need time, I think.

I mean, really, before CMS how could anyone know the forest industry knew it was creating climate change? I have thought about that a lot. I cannot see anyone knowing CMS and not caring. So denouncing wood companies isn't in my cards. What good could come from doing that? They get the first chance. And being sensitive to their economies must be practical, or we'll undoubtedly fight in the political and judicial systems and stand to lose precious time. And I do mean precious. Besides there are rules to abide.

"Those who have the gold make the rules." We all must buy in and be committed. And who is to say they won't? I think the forest industry will want to do what is right and make the money. Who wouldn't?

So, what must CMS buy? Generally speaking, CMS pays out on the net CO_2 sequestration of a forest for its past year of growth. CMS wants to make sequestration a more vital commodity traded in carbon markets like Europe. Within the U.S., that type of market is voluntary and needs CMS's permanence and support to pressure corporations to do the right thing. We're trying grassroots campaigns to reach them and having zero success. But we just started. The point to our current efforts is much of the timber infrastructure already exists, so it's not much of a stretch to have CMS working within all of it. But again, it's a money thing for sure.

MORE ON FORESTRY AND WAYS TO WIN

The real trick is to generate enough income for the owners of trees to keep them from being clear cut. If forest owners generate enough income with CMS mitigation, they will protect their new investment in *sequestration value*, their trees. And that applies to other countries as well. Especially Brazil, they have a real opportunity with CMS. All countries do regardless of economic status. CMS is a global concept without borders. Sequestration income could even stimulate forest owners everywhere to do thinning's and not clear-cutting. And use more native species to replant bare grounds. And harvest only when the plot naturally requires a thinning.

Mature trees require more space to obtain their sequestration potential. The overcrowding of forest plots today is a global epidemic. It's a major contributor to forest fires and declines in wildlife. The trees being naturally choked out of sunlight by faster-maturing trees should not be allowed to contribute to rotting CO_2 emissions, they should be harvested and used by humans-it's an available renewable that was naturally displaced, so making something from it helps everyone! Net zero products that store CO_2, like AWCs are also possible with biomass removed by thinning's.

Trees take decades to mature and develop their vital CO_2 sequestration. In return for those decades, landowners allowed to thin out the dead and dying trees to earn extra income can help provide those decades at first and then the century's needed for healthy mature forests later. Forest owners can produce *sequestration value* and biomass income from their lands, just as it should be.

Thinning's do improve the overall CO_2 sequestration ability of a forest plot by removing the dead and dying trees in timely rotations of multiple decades. But thinning costs a lot more and that's why forestry producers don't do them now. Thinning must generate revenue for the forest owners and without sequestration income they don't. But with, well, that future should arise nicely. And that brings me to the second way to win. Remember where CMS started, even before CMS was figured out? AWCs, Advanced Woody Composites.

Okay, I'm going to keep this short for a number of reasons. AWCs use forestry way more efficiently, up to 92% more efficiently, by substituting conventional lumber and lumber products that can't and don't use forestry efficiently. What makes AWCs so efficient is easy enough to explain. They use more of the tree and are less wasteful to produce compared to contemporary sawn or peeled lumber products. Think of particleboard sheets that are much more robust than plywood and made with less material. What makes them cool? Glad you asked. AWCs can be formed or machined into studs, beams, car doors, or even cups. They can be made into all sorts of strong and pliable products. Unsurprisingly, they also cost less to produce and purchase than lumber, steel, concrete, carbon fiber, etc. Amazing, right? Yes, they are, but it's not easy to get AWCs off the ground to begin curing climate change.

Out of CMS and AWCs, CMS offsets from its stewardship of forests are the least expensive and most CO_2-sequestering kind of climate mitigation. CMS is much faster at curing climate change than AWCs can currently be. Over time, the two programs accomplish the same climate objectives, maturing trees. They can do even better in consolidated ways; however, we need a speedier cure first. And we need it right now. That speedier cure is CMS.

CMS stewardship makes an instant difference: mix money with trees and let them mature! On the other hand, AWCs could take decades and heaps of money. I don't want to get into the details and blur the message, so I'll try and keep it simple. Did I mention burning vast piles of cash long before they have influence in climate change? And that pile of burning cash would not return any sequestration value for a decade. Too much time and money makes AWCs risky and slow to act when compared to CMS stewardships.

CMS mitigation provides a full toolbox of opportunity to fix it all to include AWC's later. With CMS stewardship, we buy and monitor trees. Easy-peasy, right? It's not that simple, but I wish it were! No doubt, it's less complex and easier to manage than AWCs. Obviously, AWCs can seriously impact climate change, but you have to be able to sell millions of them regularly. This might not be a problem given that they're less expensive to buy and ship than many current wood products. But again, it could take decades to get them working towards fixing climate change like CMS can do right now. Either or both are legitimate climate winners because they both address sequestration, but CMS stewardship is the much lower-hanging "climate-cured" fruit. And absolutely, I'd love to do them both. But I'm one mad scientist and a few overworked volunteers so we must pick our priorities and CMS is it. As usual, it's really all about the budget.

CMS ECONOMICS FOR THE FOREST

Trees are a resource for humans to better themselves. CMS doesn't want anyone to stop making a living from forestry, we do want to modify how they do it. Change is always a tough sell. Add together contemporary forestry economics and the demands of naturalists like me, and undoubtedly, the relationships needed to make CMS work can and have been challenging. Logically, you would think the study's new knowledge would trump any challenges in relations. It does not, because of the economics involved.

Economics are the key to making CMS work. CMS wants the wood industry to become efficient with the renewable but *binary restricted resources* they manage. That focus of the study is completed by CMS economic stewardship. CMS increases the value of standing forestry for its owners in line with the world's climate needs. The idea is the world, naturalists, and tree owners divvy up the relational win. CMS will end negative aspects like clear-cutting immature trees and add native species back in, improve biodiversity, and give our furry little forest friends places to earn their living and create more furry little fellows. Plus, it helps eliminate forest fires, improves watersheds, puts the beetles back where they belong, and provides a lot of fresh air. Forest owners earn money to make it happen because nothing is free, and it is their tree. And city folk outnumber the people who'll do the work 10,000 to 1. So, it's the city who needs to buy in more. But the world's animal populations are the largest winner as sequestration improves with each passing year allowing climate change to quickly dissipate into a bad memory.

It was no secret before CMS that healthy forest ecosystems support healthy human societies. That has been a given since human time began, or our predecessors would not have been nomadic and changed zip codes to a better forest! Focusing on the study's knowledge influences potential CMS sequestration producers into climate solution decisions. They are human and just like us. Really, they have the same bet on the climate change table and just as much to lose because there is no tomorrow in our shared predicament. Nonetheless, we will remind them every chance we get and start with the following:

The CMS study expanded, defined better, and restores the hundreds of millions of years of a previously magnificent symbiotic relationship humans had with forestry. Restoring that relationship with free market economics is forestry stewardship at its best: being in service to human domestication's *sequestration value* and answering *forestry demand* with efficiency, that is Complete Mitigation Science's game plan.

Plus, the end of clear-cuts and allowing trees to mature naturally forever are certainly great positives. Oh, saving the world from climate change isn't a bad accomplishment either. All that is worth CMS compromising between views of fat-cat industrialists' and the tie-dye-wearing naturalists.' Only now, CMS can put those two together by paying the one providing the values of the other a place to grow.

But if you don't like it, don't get in the way, because there's a plan B, C, and X.

SOMETHING FOR EVERYONE

When the captain of the Titanic was informed by his radioperson, "there could be icebergs ahead," he thought about it but still said, "Full speed ahead, I have a record to beat!" Blame it on peer pressure or whatever you want. He made the decision to hit the iceberg. Some foresters will be just as happy to hit the same kind of thing. That we are assured of because some people you just cannot help in anyway.

CMS fully expects to deal with the "in charge" who'll be screaming, "Full speed ahead, climate change doesn't exist, wood is renewable, it's green tech," or the most uneducated one possible, "humans don't need to do a thing!" Being devoid of science in any decision processes opens one to believing and fighting for irreputable nonsense. In which case, they will hit an iceberg made of CO_2 (dry ice is actually made out of CO_2, ironic isn't it?) After the crash, they'll make their uninformed decision much worse.

We expect that the CO_2 iceberg the study made for them to wake them up. But hey, we aren't ignorant to the plague made by egos. SO, we understand, some will remain comatose to CMS knowledge and will continue to scream, "Full speed ahead!" while the world continues to sink towards climate chaos.

It would be better to avoid climate chaos, so let's sink their corner of the market or pass their material supply to any competitors who are willing to act on the world's behalf. At least that is plan B. CMS won't be suffering fools. I've told you my plan; I'm going to beat climate change with CMS or go down fighting, I meant it. This is why CMS is currently a business venture. First, so we don't have to suffer fools or their influence. Second, it's the only legal way to spread money to landowners and all who help CMS succeed.

Essentially, we want to make CMS part of the public realm, an investment so sequestration value can perpetuate forever and continue to mature trees indefinitely. We think that is only possible by establishing sequestration values in global cap and trade and in a managing company participating in free market enterprises. To be sure, it is all unproven management concepts currently being evaluated by us. And so far, our concepts have been definitely more out the box than in it; sort of like CMS itself. The reality is that only time will tell if we got it right or have to adjust in the future. So, what structure CMS takes on in the future to manage its affairs is debatable. Who knows, CMS could end up a tax-deductible no-profit at some point, but we like the idea that anyone with a stock certificate can benefit and help perpetuate CMS mitigation. Hmm? While we get that all worked out Engineered Wood Company funds CMS, EWC. They are a "C" Corp and doing the best they can with what they have. I also like the people involved with EWC a lot. But here is something you should be aware of, post CMS knowledge.

In these last weeks of writing, I have seen one of the largest, most profitable timber baron companies advertising how green they are with their renewable wood products. Oh wait, that's all of them on that bandwagon. What I meant to mention is how a particular state forestry agency is advertising how green they are. Oh wait, that's all of them again! And again, they were unaware of CMS when they produced the following:

When you see the advertisement that says, "Hey, look at us, we make 'renewable' wood products," or my favorite, "We plant three trees for every one we cut down," you now know better. But keep in mind, they might not. Not yet. It's more than possible that they haven't even heard of CMS. Maybe you should inform them with my complements and my reference; but please be super nice, they really didn't know. Anyway, just in case you need it, here are a few points that make a fact-driven argument countering those "renewable" forest advertisements.

1. They have to plant 3-5 saplings (trees) for every tree cut down because of the attrition rate of sapling replanting. Of the 3-5 saplings replanted, 2-4 will die during the first few years, leaving only one. Perpetuating not improving.
2. Sequestration is not as renewable as biomass, the *binary restricted resource*, so we'd be better off with the one tree they cut down or the forest they clear-cut.
3. Wood, biomass is renewable, but the *carbon hump* from clear-cutting ensures its use is not a net CO_2 neutral or negative even after replanting. It is a CO_2 emitter and taking that into account also makes the wood products CO_2 positive.
4. Only 20% of all the wood harvested remains carbon neutral or negative. The rest is part of the *carbon hump* and releases the remainder of CO_2 from the harvested tree within 10-20 years after harvest. Yes, that includes materials like 2x4's in buildings and toilet paper.
5. Finally, they are perpetuating climate change by impeding CO_2 sequestration. Just ask them how old that one

tree they cut down was and if it was harvested in a clear-cut.

Seeing all that and seeing those advertisements provides a new perspective on what *constrained deforestation* is and how it causes climate change with modification of the human biome. Yes, it is geo-engineering we must undue by working with "them."

Now for a real shocker:

Ask them who paid for the replanting. More often than not, it's not or not entirely the timber baron paying the bill, if at all. They prefer natural regeneration, basically doing nothing at zero expense by allowing the logged area to reseed from adjacent trees.

Now onto who owns the land that harvested tree came from? If they answer they do, then ask how long they've had it and who they got it from (which should all be public record). When big timber does replant trees, they love to tell everyone about it, hence the advertisements. In truth, they let a no-profit replant their land so that it can be harvested again. Yeah, that's how the scheme works. The no-profit publicly sells themselves to gain donations, then allows the government to pay for some or all of the replanting cost, and then they sell the land back to the timber company to harvest. Total fraud on the donors'? I am not an attorney so I could not really answer with expertise. But I am ethical so yes, yes, it is. But ask yourself, "is replanting possible without a scheme like this...in any other way?" I cannot help but wonder because it is super expensive to replant. And it's almost impossible to sell the immature trees for enough money to cover it.

So here are more questions to asked when you're going to get involved in any "tree planting" promotions. Where are they going to be planted, who's planting them, and how long will they be allowed to grow? And who's going to take care of them while they grow? Finally, are there any timber companies or no-profits involved? If there are, well, honesty is called for.

To finish this section off I'm going to point out what is important to CMS success. Those in charge need to be kept in profit. It's the first line of their job description. MAKE MONEY! CMS will go out of its way to make it profitable for them at every opportunity. Now, CMS expects forestry growers lined up out the door the minute CMS can start paying for sequestration. I really don't think that's much of an imagination stretch: money and saving the world seem like good motivators. But I also thought Google was a sound investment. Oh, wait, it was. So yeah, money is a pretty decent motivator to forestry businesses.

So now all we need are trees, cooperation, grants, donations, carbon credit sales, government support, another book, videos, an XPRIZE stage, cap-and-trade accountability, standards, internet gurus, historians, atmospheric scientists, forestry... I will stop, we need you. The list goes on forever. But do you know what? It is all doable.

How do you eat an elephant? Take one bite at a time while hoping it doesn't rot before you finish. But it sure helps if volunteers show up with their grassroots BBQ sauce. And no, I would not eat an elephant for real. They are an endangered species that CMS forestry stewardship will help as well.

NOW, SOMETHING FOR EVERYONE

The truth I believe in is that people are intelligent, although I've been proven wrong before and I know better than to actually count on that being true. The thing is, it's never good to bet against people, ever. So why do I have such faith in all folks? Because I've witnessed and been a part of groups of people who get things done and make things better. Democracy really works and is statistically proven. Because, individually, most people are way smarter than most politicians and media outlets give them credit for. Especially, when tele prompted mouths provide only opinion and never enough facts to make good decisions.

- Our intellects are dying on the vine from an absence of unbiased facts. Fact is being removed while truth is being interpreted for us in order to create diversion from unbiased truth. Climate change opinion in the public is no different. CMS needs fact-based opinion to succeed.

Okay then, the truth as I really see it, our societies aren't reasonable and intelligent because some of us accept tainted opinions much too quickly. And unfortunately, there are tests that confirm many societies are totally underinformed by their own uneducated opinion. Which makes useful information a rarity these days.

But beautiful and really significant things can happen when people pay attention and act within their groups. I'm talking about the "grassroots" thing of spreading information that can make real differences. And it has and does work to educate as well. So where am I going with this?

It all relates to the beginning chapters of this book: "But Will Humans Use It?" I truly fear we will not. Right now, the fight to persuade CMS knowledge to the public and then into the one's in charge looks unwinnable to budget and time left on this Earth. I am not giving up; but then again, I am not going to beat a dead horse either.

If I cannot get advocacy for the study, I'm going to chalk it up to no good deed goes unpunished and go enjoy my last 15-20 years with the wife. To hell with anyone who are too important to bother to prevent their coming demise. A demise that I will not live long enough to experience. I at least have provided an alternative to extinction, if folks do not like or refuse to even try and understand CMS, well, I have done my part and provided the nicest warning and timeline possible.

That last part, "the warning," I should mention my history indicates my warnings should be heeded. To date, I've forecasted a lot of train crashes to people and organizations only to laugh at or take advantage of them when 100% of those crashes occurred. And they always occurred because I do not forecast unless I can also prove it. So be my guest, ignore me. I for sure will not be around in thirty years and by ignoring CMS knowledge everybody will likely be on the fast track to joining me, and that statement is regardless of somebody's age today. No one can hide from the runaway CO_2 levels caused by lack of or the ongoing impeding of sequestration. Nor will they experience current climate predictions because those are devoid of sequestration, a new knowledge on the other side of the climate equations.

I do not like being angry and aggressive. Please consider me enthusiastic about CMS and forgive my Scottish Thrawn. I really dislike the drama the study has caused in climate knowledge. To a one plus one guy like me that drama now and in the future is regrettable. So, let's go a different route offered by nonscience experts. They've told me we need an army of ants, so let's get out the sugar.

Here is the problem defined to me by more than a few marketing experts telling me what I already know. "Intelligent people allow society's interference to become a reasonable assumption. They do so just to get along, and to survive this toilets flush to the next day." That interference is ads, click bait, and algorithms socially engineered so humans don't have to think about whatever it is. "They" answer for us and we accept "their" answers because "they" are "influencers." Personally, history tells me that influencer thing is a house of cards. It makes things happen here and now but really lacks the longevity facts do. Unfortunately, our world is full of influencers who readily sell their credibility to entice us into believing misinformation or into buying something right then that we don't really need. All humans are potential prey to these predators acting like animals. Snake Oil Sales Inc. Step right up for this fantastic product! I use it myself and it's great!

Now, don't be shocked by what I write next. CMS is going to do all that as well. We have to, OR I should say all the questionable people and companies controlling information these days are making us do all that. Still, CMS knowledge needs to spread and doing that requires it to be seen and heard.

Well, it's not dropping our ethical practices or lowering our integrity. It's more like equalizing with societies. We don't need loners (like me), we need a collective society, a cast of millions, and all that click bait, bandwagon stuff works to get information out. Okay, we won't reach the people we need to, not at first. We'll mostly reach free riders who won't lift a finger to help CMS. But it's a start, and the people we contact might help us reach people who will act. The CMS grassroots movement will hopefully grow like a sunrise. But hell. Who knows what will really happen, probably the only thing that will grow is my disappointment. Will humans use it?

I will say that we are way different from other information campaigns I've seen. The difference from the typical junk we see and hear daily is made from CMS's credibility. I think of CMS advertisement as an information commercial that's useful and not governed by sales numbers, its governed by people being made aware. We will use facts and cure climate change instead of tracking some sales conversion to the dollars spent. Our media mission is to inform, to implement, and not necessarily profit. Don't get me wrong, we need a little profit to keep the workers from revolt, tree growers paid up, and toner in the printer.

NOW, SOMETHING FOR EVERYONE FROM LITTLE OLE ME, PERSONALLY.

To all those really exceptional people that make up an exceedingly small population in the United States of America. If you are not seeing the CMS picture by now, you never will. Congratulations your now a well proven moron. I also regret to inform you; you are on the fast track to extinction. That is if a self-inflicted act of Darwinism does not get you first. Please find a cave somewhere and stop bothering good people with your click baited bot and AI generated bull puckery. Our society can only hope for a legal cure for you! Oh, and please don't breed, that only makes it worse for the rest of us.

All joking aside, the CMS study is not a joke, nor an attempt at some misguiding fraud or make-believe, although those are the receptions I regularly get, and I do mind. But, please, do not compare me to the people in the first paragraph. However, that sort of negative stuff has declined, and it really has been more positive receptions lately. The longer CMS is out there and studied the better the reception I get. However, negative CMS commentary will never completely go away because its popularity growing means it will be nefariously placed onto even bigger stages by the zillions of click baiters who can profit by selling hate, discontent, and conflict. For the record, I like a lot about what social media has done for communications, staying connected. I hate that tick turds can profit just by getting someone to click on their site. Oh, where was I, oh yeah, getting picked on regularly, initially.

    SO, I've been called a fix-climate-change nut, a tree hugger, just another crazy person, and a lot worse! Well, I'm not a tree-loving nut-job. I'm just a guy who's scientific obsessions turned up something unusual and existentially inconvenient to hear or promote. The study results are really the mother of all inconvenient truths. -I kind of quoted Al Gore's book, I hope you don't mind. I kind of do because I read his book and I'm not sure it wasn't written in advocacy to anything but personal profit. But what do I know? Anyway...

- 🍃 Should you decide to contribute time and or money to CMS, you should know everything.

I want nothing to do with any of this. No kidding. It is a responsibility that I unknowingly walked my science obsession into. I did not go out searching for Complete Mitigation Science. I had no "clear" idea what the study would reveal, and I didn't think it would be the cure for climate change. Therefore, I was twice removed from knowing CMS would need to be promoted! Let me explain just how dumb I can get by being a fact-based person.

After seeing the study's conclusions, I thought just showing it to people would generate a new level of enthusiasm, a big win. Man was that thinking naïve. I was so excited, the study's proofs were easy peasy to understand, it is real and reproducible even by most anyone. I really thought an actual cure for climate change that works would excite folks. It does not, because nobody believes it, not initially. Eventually, yes people do accept it, it takes time. That experience beat the hell out of me with a club made of patience. Plus, that first attempt at announcing CMS incited my own worry to "will we use it?" The experience also makes me reluctant to be the CMS spokesperson. Because you get garbage thrown at you first and a long time before the punchline actually takes effect. It can be weeks or even months before the receiver and logic mix.

I do not want or desire to be CMS's spokesperson. I am doing what I can do, hopefully well, until someone else can pick up the flag and do it better. In many CMS tasks that is not really asking for much at all, because I suck at a lot of the tasks needing done. Accordingly, I do this job the best I can. Only because I'm personally responsible for the mess CMS has to make. It's difficult for me to ask someone else to take the initial ridicule I am responsible for. I am accountable for CMS's future. It darn sure isn't for the riches. CMS has to borrow just to buy paper clips and hot water in the bathroom. That reminds me; it's my week to supply the toilet paper. For now, responsibility has its rewards and really long lists of punishments to prescribe.

My responsibility really begin after defining the logic and writing out the study's conclusions. Prior to that, I enjoyed the research reading, designing parts of the study, modeling as the study found questions, and figuring out how to perform experiments, and eventually testing proofs! I was tinkering at first and I enjoyed that. The deep dive put an end to the tinkering and years later the business tasks surfaced and became more apparent and in complete disregard for my strengths. The to do lists had begun and become longer by the week.

It started with "make sure people become aware." We're trying! Then it became, "Let's get CMS up and running and I mean now!" And let me add up all my costs. -yikes! Now, I have gone and authored this book, and other writings are soon to come. Oh yeah, the web site blog, I keep forgetting! And then, the meetings, more perception ridicule at my expense, learning curves for internet black magic, and all with the pressure provided daily by minimally funded activities actually waiting for funding. If you can explain that last one your hired! As a result of these undesired pressures, I thrive... Okay, Okay. I just needed to stop the whining. I'm getting crushed. It sucks, but I kind of like the way it sucks.

SO, I don't have any real complaints, just the normal at work gripes we all share. We are dining on Elephantidae after all. Many really good things are occurring as well. In fact, so many that I'm encouraged almost weekly. Stuff like the CMS/ EWC web site hit 6,800 visits today and all within three months! So yeah, it's a good start. Hopefully, some will come back and help out. Our list of supporters/subscribers increases almost daily. We're making headway on the XPrize work as well. Lots to be grateful for! And I am.

I'm really grateful for EWC's people funding all this. They do what they can but we're talking paper clips, printer toner, and reams of paper, not tree leases, land purchases, or travel expenses. This is all a lot of work because of the lack of money and it's making me older faster than I prefer. But I do like a good challenge. It is a challenge because I'm not a writer, salesperson, or internet guru. And now, I'm told I should make You tube videos. It never ends and I don't expect it to change, ever. So, I will give it my all, but not because I want to. It is because I must. Just like you! And why should we bother?

Well, now that CMS defined climate change, fighting climate is easier and less costly than it's ever been. So, what's stopping us? Well, that answer is this book's entire message, *you*. Yeah, *you*. Nobody is going to do it for *you,* and I can't do it without *you*. This book, Engineered Wood Company, and the study's knowledge provide opportunities and invite *you* to help CMS save the world. Please use them.

TIME TO CLOSE UP... FOR NOW

And yes, the goal of this book is to get you to contribute to CMS in any way, shape, or form possible. And your help does not require your money. I mean, I hope you paid for this book because that's a great start! But we could use more of you if you could swing it.

I am CMSing, and our frail and broken planet needs you CMSing as well. Just getting started sucks. Things will progress and change as CMS continues to move forward. SO, a lot of what I say below is still working itself out and could use someone with your experience and talents.

Just how do I help? I'm so happy you asked!

- Join the team! Misery loves company! Send your what I can do stuff!
- Subscribe for free on the web site and stay informed with our newsletter.
- Volunteer with CMS! Contact us through the web site and send a resume'!
- Contact us with some feedback and your suggestions!
- Link your web or social media to the CMS web site, help spread the cure for climate change, CMS!
- Make a YouTube video and link it to our web site or Pay Pal donation page!

- 🍃 Yeah, social media, do that!
- 🍃 Help endorse CMS with show and tell! Tell anyone who'll listen! Educate! But duck and forgive when they initially throw the garbage at you. They'll eventually get it, let logic soak in.
- 🍃 Send our web site link to your contacts. Sign your email with our link! We're going to put a downloadable green earth CMS logo on the web site soon.
- 🍃 Recommend this book to people if you liked it. If you didn't, still recommend it, with your caveats!
- 🍃 Donate to CMS one time or become a monthly contributor! Pay Pal awaits! And anything helps.
- 🍃 And the most help you could provide is purchase or even sell $CMSco_2t$ carbon credits. Be a part of the growing volunteers in promoting Cap and Trade by becoming a member of the CMS's Carbon Exchange beta test! Have a Carbon Party! Invite Friends! Mature Trees! And save the world!

If you want to learn more about CMS and AWCs, dive into the hundreds of details this book doesn't specifically mention. Get the *Complete Mitigation of Atmospheric CO_2 and Emissions, Complete Mitigation Science, CMS* study manuscript and my notebook and get after it. The note's book is all my notes, models, cited works, internal study's, and other findings-good and bad. Oh, and it has tons of pictures! No, not really but there is a lot of graphing. It's not available at the date of this writing, but it is coming, soon? I really do not know when. There are also CMS journal manuscripts. Once published, they'll be free, but they'll have to be licensed for any commercial use by the publisher and the copyright owner EWC.

I know you know EWC stands for Engineered Wood Company. They are private investors, and me. They funded AWCs and 100% of CMS's work. All of us are now putting CMS mitigation plans in place, implementing. That task gets bigger every day and CMS needs it globally implemented. That is the goal.

I more or less run the CMS side of things and I am struggling from lack of help; expertise is what I should have said. But overall, it has been a great relationship with EWC that's brought us here and has led to bigger plans for CMS's mitigation. EWC is the company trying to change the world with CMS one tree branch at a time. And, so far so good, so CMS is staying put.

- The video thing on YouTube, I don't, aagh! Help! Any volunteers?
- https://ewc.company is the website for CMS, for now and maybe forever.
- If you can help in other ways, registration on the website is the way to get it started. Use the contact

and about section. And tell us what you can do! One thing is for sure: CMS needs all kinds of help, oh, did I mention that already?

- Educators, and STEM folks, more is coming on an educational website. That's in the budget. But until I find someone with appropriate classroom experience and not a horrible and scary ogre like I am, you're stuck with me, and my apologies for being me.

Finally, all proceeds from this book, donations, or anything CMS-related go towards implementing CMS, promoting CMS, and continuing to educate the world about CMS; so, others can fight climate change with us!

Oh, look here, that sneaky Figure S1, where climate change began and where we end! Thank you for reading. Opps! I meant to say be sure to read the glossary if you have not already. It was long months of arduous work defining the logic, the least you could do is read about it in the glossary! Thanks again and I apologize for being me!

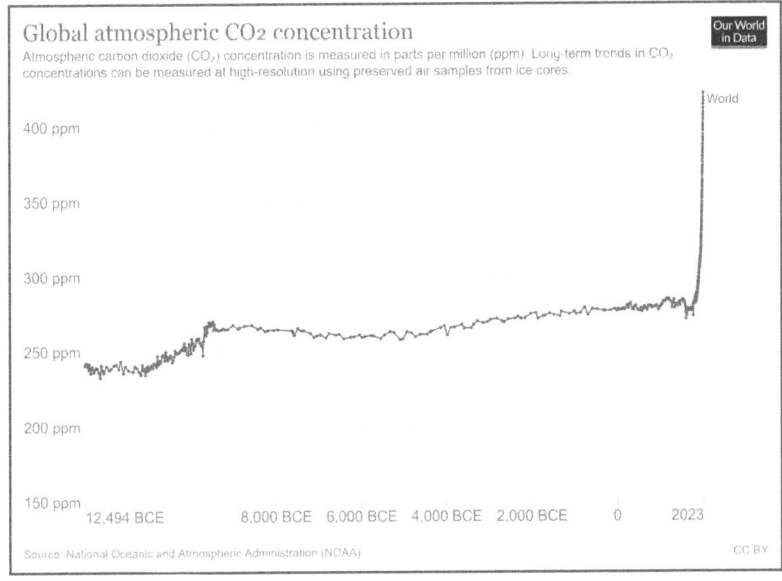

Supplemental Figure 1, NOAA's data, not mine.

The upward CO2 ppm trend starts around 5,000 BCE and continues to around the 1850 datum as the beginning of the climate collapse and then to the 1950's when the runaway effect starts. It all correlates to *demand driven forestry's* history that was geoengineered for human domestication.

The upward trend in ppm began around 7,000 years ago, around 5,000 BCE. It did so with increasing human populations creating more forestry demand. That demand evolved into *demand driven forestry practices* which first eliminated mature trees and replaced them with numerous immature trees. Doing so also implemented *constrained deforestation* by harvesting the smaller and smaller trees sooner and sooner. Eventually, after 7,000 years, no global forests are allowed to mature. Hence, a tree's *binary restricted resource,* sequestration, is physically impeded to net carbon positives because trees are harvested at younger ages each year. The results creates an environmental impact that is 100% reversable with the restoration of sequestration with tree maturity.

GLOSSARY

Terms "Complete Mitigation Science" coined or expanded to describe climate change.

1. Sequestration value

1.1. Forestry CO_2 sinks and their sequestration ability are more nonrenewable than renewable (see *binary restricted resource*). This generates value in climate mitigation regarding carbon credits, offsets, and is bolstered by *sequestration dependence*. Herein, the ability to sequester CO_2 within climate mitigation is described as a human endeavor toward domestication, which can have an economic value. S*equestration value* is established by considering CO_2 sequestration abilities within terrestrial sinks with human domestication goals governing valuation. *Sequestration value* is currently priceless, figuratively, but can be valued socially, within exosystemic approaches, and also appraised and monetized.

1.2. CO_2 *sequestration value* is established by measuring the requirement to maintain atmospheric CO_2 and abate climate change. It can be balanced with emission input, more abundant than emissions, or placed in deficit to a desired level, as it is globally today. The current CO_2 levels within the atmosphere are notable. (1) Therefore, removing CO_2 to achieve balance, abundance, or deficit creates a high sequestration value socially and economically. *Sequestration value* can be monetarily established by converting atomic mass units of CO_2 (44 amu) as the *sequestration value* or the sequestered amount as carbon (12 amu) stored and as the result of sequestration. (2) (3) (4) (5) Thus, *sequestration value* is an amount of sequestration achievable or having occurred within sequestration-based climate mitigation that has been or can be further translated into economics to facilitate ongoing human domestication.

2. *Binary restricted resource* – an expanded definition of a renewable resource

    2.1. Trees are renewable resources. Biomass from trees is one part of this resource. Another part is a tree's ability to sequester CO_2 (serving as a terrestrial CO_2 sink). The latter of which is less renewable than the former.

    2.2. Trees replenish biomass over two to three decades. However, the duration required to become a viable CO_2 sink and contribute to Earth's biome as a net CO_2 negative is a much longer duration. Therefore, *sequestration value* of a tree cannot be as easily renewed as its biomass (1). In contrast, the tree's *sequestration value* is more apparent to human domestication requirements than demand for its biomass.

    2.3. The study's findings recommend the tree or forest as a CO_2 sink is more essential than the biomass among the tree's combined resources. However, unlike its biomass counterpart, the sink resembles a nonrenewable because of the extended time durations to renew its s*equestration value*. It can take a century for the sequestration resource within the tree to replenish. Thus, although renewable, its efficacy is different from that of its biomass resource.

2.4. The demand for biomass replenishment necessitates a much shorter renewal duration, significantly suppressing the extended duration requirement of the tree reestablishing as a CO_2 sink and the *carbon hump* it also contends with. Thus, restriction is applied physically to the sequestration of a tree by the demand for its alternate resource that requires a much shorter duration.

3. *Unconstrained and constrained deforestation practices*

The scope of "deforestation" is broadened into two factors with differing roles. These assessments oppose the conventional definition of "forest degradation," whereas forestry use is expected to eventually result in afforestation with an anticipated return to a normal forestry state.

3.1. *Unconstrained deforestation*

3.1.1. *Unconstrained deforestation* permanently reduces or eliminates the binary restricted resource of CO_2 sequestration by substituting land use or clearing forestry CO_2 sinks from existence. Like deforestation, clearing forests for crop production or an unintended demise from weather, fire, climate change, biome formation, urbanization, or biological effects are classified as *unconstrained deforestation.* The concept behind *unconstrained deforestation* is that the forest will never return to normal forestry.

3.1.2. In contrast, deforestation is the removal of forests but implies afforestation as a possibility. Thus, the problem is defined as being related to the inherent and enduring characteristics, such as those of urbanization, dam construction, and bug kills which continue to be deforested despite the natural worlds attempts to regenerate or alter the previous displacement of forestry. Therefore, results are unconstrained from perpetuating.

3.1.3. *Unconstrained deforestation* is also responsible for increasing unintended climate demise in an ever-expanding spiral. The more *unconstrained deforestation* occurs, the more climate degradation, which leads to biological biome changes, forest fires, and adverse weather conditions, which in turn results in further *unconstrained deforestation.* In the end, all these factors are adjusted via climate change impacts. The action is also unconstrained in creating the destructive circumstance and perpetuating its result.

3.2. Constrained deforestation

3.2.1. *Constrained deforestation* is the forestry stewardship that allows forestry to be cultivated but restricted to immature growth and never allowed to recover its *binary restricted resource* sequestration. CO_2 sequestration is not currently considered in forest practices regarding renewability. This effect is typical because of human interference with natural biology with current and generationally applied stewardship practices. Those forestry practices comprise short-durational commercial harvest rotations that reduce tree maturity, lessen biodiversity, and increase forest impediments due to other recurring impacts like forest fires, bug kills, or drought.

3.2.2. *Constrained deforestation* can be summarized as "the unnatural and inefficient use of forestry that impedes the capability and quantities of CO_2 sequestration resulting in *impeded fast-cycle sinks.*" *Constrained deforestation* practices also emit more CO_2 from forestry than the replanted sinks can sequester, owing to human demands for unnatural and inefficient uses that create *carbon humps*. Managing forests for product demand and not regarding its *binary restrict resource* is crucial to the term, making *constrained deforestation* a precedent within this study. The action is thus constrained by the destructive circumstances perpetuating its results.

4. *Climate change datum*, beginning of climate-changing results.

4.1. The datum point of CMS is a timeline indicator at approximately 1850 CE. Precedence is established using four indicators: 1. human and natural CO_2 emissions accelerate an existing upward trend in CO_2 ppm atmospheric levels; 2. The shrinking volume of existing forestry fast-cycle sinks abilities to sequester CO_2; 3. the downward trend in usable forestry CO_2 sinks available, primarily due to the *carbon hump*; 4. The correlations of atmospheric CO_2 ppm data to historically documented *constrained and unconstrained deforestation* practices and their expansion globally.

4.2. The datum marks a pinnacle of traverse starting in the period of the last mature forests on Earth, located in the Americas (North, South, and Central), except for the untouched portion of the Amazon Forest, becoming marketable and subject to harvest from the 1700s to present. As such, the datum is notable within the correlation between historical forestry use and the acceleration of atmospheric carbon dioxide levels after 1850 CE.

5. Impeded fast-cycle CO_2 sink or impeded sink.

5.1. Forestry or other *fast-cycle CO_2 sinks* that exist as immature within the *binary restricted resource* definition as their CO_2 sequestration ability is impeded from reaching its higher volume potential. As opposed to being mature, *sequestration valued,* or as sinks that are functioning unconstrained by humans. Typically, *impeded fast-cycle* forestry sinks are impeded by *constrained or unconstrained deforestation*. The result restricts higher volumes of carbon dioxide sequestration from photosynthesis by limiting maturity or existence.

5.2. An impeded sinks magnitude is measured in metric tonnes of carbon dioxide annually sequestered. The magnitude is obtained by comparison of having a human absence in contrast to any current or previous harvesting that occurred. The solution is given in percentage to what exists currently to what could have with human absence. The *carbon hump* should also be considered in the percent impeded evaluation.

5.3. Thus, the eliminating, impeding, or decreasing the volume of a CO_2 sink's sequestering ability is obtained by it's past, present, or projected levels into a current evaluation if it had not been harvested.

5.4. *Impeded sinks* and impeded sequestration is *proportional* to tree or forest maturity, complete climate mitigation potentials, and measurement of sequestration-based climate mitigation.

6. Law of conservation

6.1. The *law of conservation* is used within the study to demonstrate the carbon dioxide amount placed and managed within an enclosed system. Thus, establishing the same condition and constraints of CO_2's presence within Earth's biome. (6)

6.2. Carbon quantities do not and cannot increase or decrease in any way within that enclosed system (Earth-bound); an increase is impossible without an external addition to the system. For instance, an asteroid hitting Earth. Therefore, Earth-bound systems producing or reducing CO_2 conform to a *law of conservation* that acts as a restriction. Overall, there can never be more or less of an element within a closed system like Earth. Nevertheless, an element can change forms, like elemental carbon, transforming into carbon dioxide, which can then move to a different location in a chemical reaction, such as photosynthesis, converting CO_2 into a C containing biomass.

6.3. One focus of the study; the physical movement of elements in an enclosed system. The evidence suggests that the conversion of carbon through human domestication requires an equalizing transfer to maintain climate homeostasis. As a result, emissions, and storage (conversion, sequestration, and sequestered) play a crucial role in achieving homeostasis within a closed biome such as Earth. The study establishes the constraints of the *conservation law* as the amount of CO_2 produced by domestication, which is found infinite in volume and unavoidable in domestication. In contrast, reduction as molecular conversion (sequestration and storage) is finite when influenced negatively by ignoring the *law of conservation*. Thus, not adjusting emissions and/or sequestration to equalization during a basic analysis within a chemical reaction or equation is problematic.

6.4. In the study's precedence, a sequestration increase is found wanted on one side of a climate equation to equalize emissions levels. Sequestration is required more so than emission reductions because the sheer volume of natural CO_2 emissions (as much as 10 times more than human) must be accounted within mankind's inability to influence that volume. This analysis accounts for the *law of conservation* with equalization within the biome, as in never more or less of any given element is possible and is found unbalanced because of conversion of the element Carbon into molecule form of CO_2 and then back to carbon within sequestered storage.

6.5. Furthermore, mitigation efforts that do not involve sequestration as in systems that discount the significance of CO_2 conversion into carbon storage and known as emission reduction or elimination attempts are all reasoned to be destined to fail climate mitigation. These attempts violate the *law of conservation* by only partly addressing the smaller side of the of the total equation within human control: human emissions. Therefore, emission-based attempts cannot and do not address the estimated 400-750 giga-tonnes of naturally occurring CO_2 emissions to C conversions required by current CO_2 driven climate mitigation. Thus, implying the *sequestration dependency* of humans and CO_2 driven climate mitigation.

7. Sequestration and emission-dependent

7.1. CO_2 emissions are defined as unavoidable in achieving domestication goals, and therefore, humans are *emission dependent*. In balance, humans also are *sequestration dependent*.

7.2. The study determined emissions and sequestration do require formulaic equalization for ongoing human domestication efforts in order to maintain a homeostatic biome. However, humans are highly biased towards being *sequestration dependent* because of the closed system the biome exists within, as defined by the *law of conservation*.

7.3. Unbalanced interaction with CO_2 forms and conversions allows extraneous conditions to generate undesired effects on human domestication and biome management, such as climate change, associated *tree and land degradation,* and *unconstrained and constrained deforestation.* Because humans rely on emissions, and they are unavoidable *sequestration dependent* is an implied rule of governing both human domestication and human biome efficacy. Hence, the balance of both is created therein.

8. *Application of proportionality*

Evidence suggests proportionality in defining CO_2 sequestration.

8.1. Proportionality occurs among positions within the study:

8.1.1. Tree or forest maturity is $\propto$ to sequestration ability.

8.1.1.1. Increasing forest maturity increases CO_2 sequestration ability somewhat "exponentially" given enough "t" (time) to increase the forest's age as maturity.

8.1.2. Maturity is then $\propto$ to sequestration ability and $\propto$ to atmospheric CO_2 ppm.

8.1.3. Atmospheric ppm residence time is $\propto$ to sequestration available in global forestry.

8.1.4. CO_2 residence time increases are $\propto$ to global sequestration abilities decreasing.

8.1.4.1. The more sequestration available, the less atmospheric residence time.

8.1.5. Global forestry area is $\propto$ to maturity and $\propto$ to atmospheric ppm reductions.

8.1.5.1. The older the forest, the less forested area required to mitigate climate change when measured in atmospheric carbon dioxide outflows.

9. Convenient forestry

9.1. *Convenient forestry* describes forestry resources within easy human access. Adjacent to human location, access, ample roads, railroads, low cost, lesser time required to haul resources, available labor, topographical ease to access, marketable status, and the usability of the forests biomass species make some plots of forestry more convenient than others.

9.2. In context to this study, nomadic people could not readily cross oceans to increase their forestry use. That came later in human domestication with shipbuilding. Prior to that, when nomads used up all *convenient forestry* present, they relocated; sometimes just to make the forestry resource convenient again.

9.3. All forests today are *convenient forestry* because of technology like trucks and human populations present most everywhere. In contrast, the Amazon rainforest is the minor exception. As the only old-growth forests remaining at scale, the amount of CO_2 it sequesters is Earth's only viable mechanism to slow climate collapse. This is primarily due to its untouched maturity levels and subsequent unimpeded CO_2-sequestration capacities. Therefore, much of the Amazonian Forest remains "inconvenient" forestry.

10. Tree degradation

10.1. CMS defines *tree degradation* as contributing to the forest degradation and *constrained deforestation's* effect on biomes. Tree degradation is brought on by *demand-driven forestry* practices. Specifically, harvesting smaller trees within a reduced duration diminishes tree maturity, biomass size, and the land's ability to regenerate naturally due to adverse biological effects and nutrient depletion. *Tree degradation* adversely affects the tree quality and, in turn, reduces the quality of wood products, precludes devaluation of tree sequestration, ultimately creating an *impeded fast-cycle CO_2 sink* and further *impeding CO_2 fast-cycle sinks* while accelerating harvested forestry land towards *unconstrained deforestation's* more permanent outcomes.

10.2. An immature tree is smaller at harvest and is less valued in terms of product efficiency but higher in value with its shortened duration for return on investment. Smaller trees lack mass according to maturity of growth and are therefore highly inefficient in processing productivity. Therefore, more trees are often harvested sooner to compensate for the inefficiency of mass lost to immaturity.

10.3. *Tree degradation* is the outcome from economically and environmentally unsustainable forestry practices. The practices are of *constrained deforestation* definition who's presence biologically oppose tree maturity and eventually degrade the forest plot used into *unconstrained deforestation* category.

10.4. *Tree degradation* increases as *demand-driven forestry* increases, propagating *tree degradation*'s effect on forestry plots by decreasing the likelihood of reestablishing a forest to its norm. Whereas a plot can no longer sustain or grow a 300-foot-tall species the plot can only grow and support the same species at 100'. The species has therefore been "*tree degraded*" to the plots degraded ability.

11. Demand-driven forestry

11.1. This is defined as forestry managed to only supply the demand for products, disregarding the sustainability of its *binary restricted resources* or the efficiency of the resource in general. Historically, *demand-driven forestry* has generated adverse biological effects with *tree degradation*. The practice is by definition *constrained deforestation*, the study's primary factor of climate change conditions forming as geoengineered. Therefore, such a forestry practice physically supplies demand and is only managed to that required extent and not to others present.

11.2. *Demand-driven forestry* is a stewardship practice that disregards human domestication criteria in exchange for meeting demand and extracting profit. (7) If economic evil exists, it will prove to be *demand-driven forestry* practices negative climate/environmental impact. Demand-driven forestry stewardship model's create and sustain *constrained deforestation's* foundation.

12. Carbon hump

12.1. After a clear cut, the grounds of a tree plot become a net CO_2 positive (a CO_2 and methane emissions source) due to remnants of the harvested trees, other plant life on the plot, or life that also succumbed to the practice that decays or is burned as biproduct. The carbon hump is added to by waste generated by harvesting and the same waste burned to produce energy or energy consumed in order to process the harvested biomass into wood products. These *constrained deforestation* practices create an obstacle to restoring any given forestry plot back into a net CO_2 negative (A CO_2 sink). The *carbon hump* is created from the inability of replacement trees as being replanted, under natural regeneration, or immature in general to sequester more CO_2 than the CO_2 positive actions (emissions) presented.

12.2. Typical *carbon humps* persist 20-30 years after a clear cut while the replacement trees mature. The study's impeded fast-cycle sink definition includes *carbon humps* to determine percentage of sequestration impediment.

BIBLIOGRAPHY

All of these papers were published by their authors for public use. So, a bit of recommended reading that I mentioned or referenced in this book. There is more cited, much more on the web site and CMS publishing. Have fun with these writings and remember who their source is! I did warn you and pointed some of the misleading data sources out in this book. I listed some here so you can see for yourself! I'll give you a hint, 1 and 9, 10-14 are great data sources. 3, 5, 6 are mostly okay and do contain some excellent data. The others 2, 4, 7, 8 all have something in common within their namesake that implies "read carefully, read in between the lines, and with great caution." Keep in mind what transparency really means, you look but you can't see it.

1. **Global sCO$_2$ Emissions from Fossil Fuels**. Our World Data, [Online] Our World Data.org, Dec 12, 2022. https://ourworld- indata.org/co2-emissions.

2. **Carbon Storage and Accumulation in United States Forest Ecosystems**. Birdsey, Richard A. Radnor : United States Department of Agriculture, Forest Service, 1992. WO-59.

3. **How to calculate the amount of CO_2 sequestered in a tree per year**, Brink. https:/www.UNM.EDU. [Online] University of UNM, 2021. [Cited: November 15, 2021.] https://www.unm. edu/~jbrink/365/Documents/Calculating_tree_carbon.pdf.

4. **Methods for Calculating Forest Ecosystems and Harvested Carbon With Standard Estimates for Forest Types of the United States**, Smith, James E., Heath, Linda S., Skog, Kenneth E., Birdsey, Richard A. Newtown Square, PA: USDA Forest Service, April, 2006. General Technical.

5. **Method for Calculating Carbon Sequestration by Trees in Urban and Suburban Settings**, U.S. Department of Energy. Washington, DC: U.S. Department of Energy, 1998.

6. **Carbon And Other Biogeochemical Cycles**, Ciais, P., C. Sabine, G. Bala, L. Bopp, V. Brovkin, J. Canadell, A. Chhabra, R. DeFries, J. Galloway, M. Heimann, C. Jones, C. Le Quéré, R.B. Myneni, S. Piao and P. Thornton Cambridge, UK, and New York: Cambridge University Press, 2013.

7. **U.S. Forest Resource Facts and Historical Trends FS1035**, United States Department of Agriculture, Forest Service. Washington DC: United States Department of Agriculture Forest. Service, 2014. FS-1035.

8. **U.S. Forest Resource Facts and Historical Trends**, United States Department of Agriculture, Forest Service.

Washington DC: United States Department of Agriculture Forest Service, 2014. FS-1035.

9. **Land use strategies to mitigate climate change in carbon dense temperate forests**, Law, Beverly E at al. (2018): 3663-3668, s.l.: Proceedings of the National Academy of Sciences of the United States of America, Vols. 115,14 . doi: 10.1073/pnas. 1720064115. See this one in acknowledgements as well.

10. **Trees Per Acre Table**, Coder, Dr. Kim D. s.l.: University of Georgia, 1996.

11. **Oosthoek, Jan K.** The Role of Wood in History. https://www.eh-resources.org/the-role-of-wood-in-world-history/. [Online] 1998.

12. **Melby, Patrick.** Insatiable Shipyards: The Impact of the Royal Navy on the World's Forests, 1200-1850. https://www.wou.edu. [Online] 2012. https://wou.edu/history/files/2015/08/Melby-Patrick.pdf. HST 499.

13. **National Association of State Foresters.** Timber Assurance. National Association of State Foresters. [Online] Dec 2021, 2021. https://www.stateforesters.org/timber-assurance/legality/forest-ownership-statistics/.

14. **Hammerschlag LLC.** Uncaptured Biogenic Emissions of BECCS Fueled by Forestry Feedstocks. On Line : Hammerschlag, NR-040(g), 2021.

www.ingramcontent.com/pod-product-compliance
Ingram Content Group UK Ltd.
Pitfield, Milton Keynes, MK11 3LW, UK
UKHW021302180426
11947UKWH00015B/968